U0227183

Wayfinding System · Public Art   Litter Bin

PUBLIC LANDSCAPE
STREET FURNITURE II

Lighting · Bus Stop · Passageway   Public Space

地景设施

高迪国际 HI-DESIGN PUBLISHING   编

王丹 李小童 刘宪瑶 陈曦 译

广西师范大学出版社
· 桂林 ·

PREFACE 序言

The term "Urban Design" has become one of those overused phrases or buzzwords that has either lost its meaning, or perhaps, never achieved the significant denotation that would help establish a common language. How did this happen? So often, we begin to categorise and label, to help us talk about concepts that aren't always that easy to understand. In so doing, the idea is changed and loses its inherent value and integrity. Such is the case with "Urban Design". We have included so many different elements and approaches in the realm of public infrastructure, that now we're not sure what it means. This is one of the reasons that so many examples of urban design have become repetitive "themes" applied to our public spaces. Rather than a physical imposition, we must respond to the surrounding architecture and landscape to sense what "belongs".

Urban Design is first and foremost about people. The growing awareness related to "Placemaking" initiatives must be centred on both the functional and aesthetic challenges to design and create meaningful and memorable public spaces where people "want to be". When a public space is thoughtfully and purposefully designed by a multidisciplinary team of artists, architects, engineers, owners, and other representative stakeholders of the community, the results can be dramatic. Without this collaboration, invariably something is missing, a connection and investment with the very people one hopes to attract and engage. Multidisciplinary design is critical to the process leading to successful and lasting urban environments.

Great projects evolve from great and compelling stories, ones that unfold in layers throughout the design of the space. Whether for the design of a public park, plaza, streetscape or transportation venue, it must be Flexible, able to function in a variety of ways and uses on different days and occasions. Community spaces must also be universally Accessible, as well as Comfortable for pedestrians throughout all seasons. Special places reflect a sensitivity to lighting, establishing tone and character, conscious of safety and ambiance as a site changes from day to night. The design of a significant public gathering space is seemingly empty without the integration of art, creating dynamic and visually Engaging sculptural installations and environments. They are rich in Detail, inviting people to discover, Remember, and return. Lastly, they must be Durable, taking into account function, material selection, and on-going maintenance. Urban design must also be thoughtful of the factors of time and scale, relevant to people whether in vehicles or on foot. Only through this carefully "orchestrated" approach to urban design, will the public be welcomed into and moved by the experience.

David B. Dahlquist
Artist, Senior Partner / RDG DAHLQUIST ART STUDIO

Dolores Silkworth
Landscape Architect, Senior Partner / RDG PLANNING AND DESIGN

"城市设计"这一术语已成了被滥用的惯用语或行话之一，或失去了自身的含义，或从未形成有意义的外延而被广为接受。怎么会这样呢？在谈到某些不太容易理解的概念时，我们往往会先对其进行分类并加以标记。这样一来，其内容就会发生变化，失去了内在价值与完整性。"城市设计"就属于这种情况。我们将太多不同的元素与方法装进公共基础设施的概念中，以至于如今我们已经说不清它到底意味着什么。这就是为何众多城市设计案例都将重复性"主题"运用到我们的公共空间的原因之一。我们必须积极地对周围的建筑与景观做出回应，去感受所"拥有的"，而不是生硬地强加。

城市设计首先要服务于大众。我们越来越意识到 "场所营造"应该以功能性与审美性为中心，从而设计并打造出人们"期望"的意义深刻且值得纪念的公共空间。当艺术家、建筑师、工程师、业主及社区利益代表等涉及多学科的团队经过深思熟虑，信念坚定地设计出某个公共空间时，其结果会令人激动不已。如果没有大家的共同合作，一定会缺少某种东西，无法吸引人也无法使人投入参与。由多学科参与的设计对于建筑过程至关重要，有助于最终形成成功与持久的城市环境。

伟大的项目源于伟大且引人入胜的故事，那些通过空间的设计层层展开的故事。无论是设计公园、广场、街景还是交通运输地点，都必须要灵活变通，兼具多种功能，而且可以在不同的时间和场合使用。此外，社区空间必须方便到达，一年四季都要使行人感觉舒适安逸。有些特殊地点从早到晚光线不断变化，考虑到安全及氛围因素，因此设计对照明有较高要求，以形成独特的色调与特征。重要的公众聚会空间的设计看似空洞，没有将艺术融合在一起，没有形成了具有活力、吸人眼球的雕塑设施与环境。然而其细节处经过精心雕琢，吸引人们去发现、铭记并再来重游。最后，公众聚会空间必须耐用、要考虑到功能、材料选择以及维护保养。同时，由于有些民众乘坐交通工具而有些步行，因此还要考虑到与其有关的时间与规模等因素。只有采用这一 "精心安排"的城市设计方案，才会吸引人们前去感受和感动。

大卫·B·达尔奎斯特
艺术家、资深合伙人 / RDG 达尔奎斯特艺术工作室

德洛利斯·斯尔科沃斯
景观建筑师、资深合伙人 / RDG 规划与设计

CONTENTS 目录

WAYFINDING SYSTEM
导视

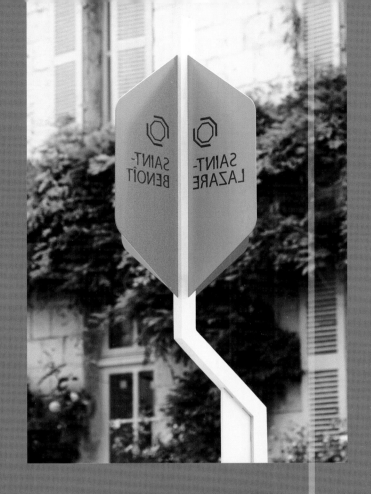

DESIGNER
Matali Crasset (Matali Crasset Productions)

PROJECT MANAGEMENT
Abbaye Royale de Fontevraud

PRODUCTION
agence pièces montées

New Signage for the Royal Fontevraud Abbey

枫岱弗洛皇家修道院的新标志

The signage project created by the designer Matali Crasset is developed around the concept of lines and beams which separate and come together again, lines like so many extensions, information, directions, at the visitor's disposal. These elements suggest staging the walk and the visit, leaving the visitor to follow some directions more than others, or to decide to take the time to read or to look for specic information.

This line, with the idea of not making the signage an invasive and essential device, appears not according to the place, but like a dotted line, moves around the site, settles on the buildings, emphasises the outlines of the passages between the areas. It lifts the walls to create a fold in which the appropriate information is available: directions, signs, the areas' identity, information on events, etc.

The materials chosen, basically in primary aluminium, with the various areas in dierent colours, allow the elements in the signage not to be in contrast with the architecture and so the intervention, similarly to highlighting, makes the directions and main information both visible and legible. Matali Crasset's line is vital as a strong creative gesture animating the contemporary city of Fontevraud, falling as such within its history, which continues to build up day by day.

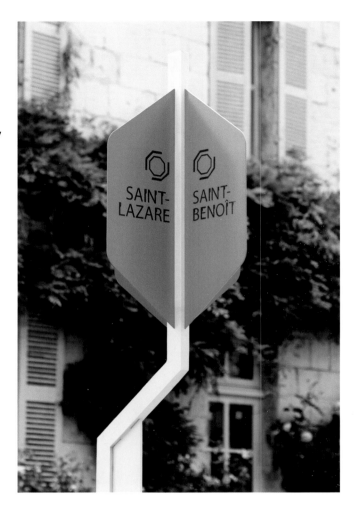

L'ORANGERIE

11h-18h

OUVERT DU MERCREDI
AU DIMANCHE

OUVERT TOUS LES JOURS
EN JUILLET ET EN AOUT

OPEN FROM WEDNESDAY
TO SUNDAY

OPEN EVERY DAY DURING
JULY AND AUGUST

ÉGLISE
ABBATIALE
ABBEY CHURCH

CLOÎTRE
CLOISTER

ABBAYE ROYALE
FONTEVRAUD

SORTIE
WAY OUT

该标志项目由设计师玛塔丽·克拉赛特设计，围绕着经纬线的分散与集合这一理念进行建造，延长线、信息及方向等许多元素都可以任由游客安排。这些元素指引人们进行散步与游览，游客可以自己跟随某些方向标识进行参观，而不是跟随其他人的脚步，或者可以让游客下决心花些时间阅读或查询详细信息。

线条的设计并不是为了使标志成为显而易见、必不可少的设施，因此它并不是依据地点设计的，它就像一条虚线，围绕着这片地方来回回回，停在了建筑上，强调着区域之间通道的轮廓。它将墙体高度提高，形成一个围栏，以提供适当的信息：方向、标志、该区域的特性、事件信息等。

所选的材料主要是生铝，不同区域采用不同的颜色，从而使标志中的元素与整体的建筑风格和谐统一。这种介入效果与强调手法相似，使得方向标志与主要信息清晰明了。玛塔丽·克拉赛特的设计理念具有重要意义，以其极具创造性的姿态，在枫戊弗洛市的历史发展过程中，赋予其生命，日复一日，年复一年。

DESIGNER
David Karásek, Radek Hegmon

DESIGN COMPANY
mmcité

OS

OS 箭头指示系统

The arrow orientation system OS applies the principle of thin direction arrows pointing to destinations and visible from all sides. Simple and effective shaping combined with well-arranged graphic elaboration of individual direction arrows represents an advanced system for guidance in all types of street environments.

OS 箭头指示系统采用了较薄的箭头设计，并且在任何方向都能看到箭头。清晰明了的外观以及每个箭头上生动独特的图案说明，OS 箭头指示系统是可以适用于任何街区环境的先进指示系统。

DESIGNER
Hans Gerber, Daniel Waeger

DESIGN COMPANY
Minale Tattersfield

CREATIVE DIRECTOR
Hans Gerber

CLIENT
Sydney Harbour Foreshore Authority

Darling Harbour

达令港

LOCATION
Sydney, NSW, Australia

PHOTOGRAPHER
Minale Tattersfield

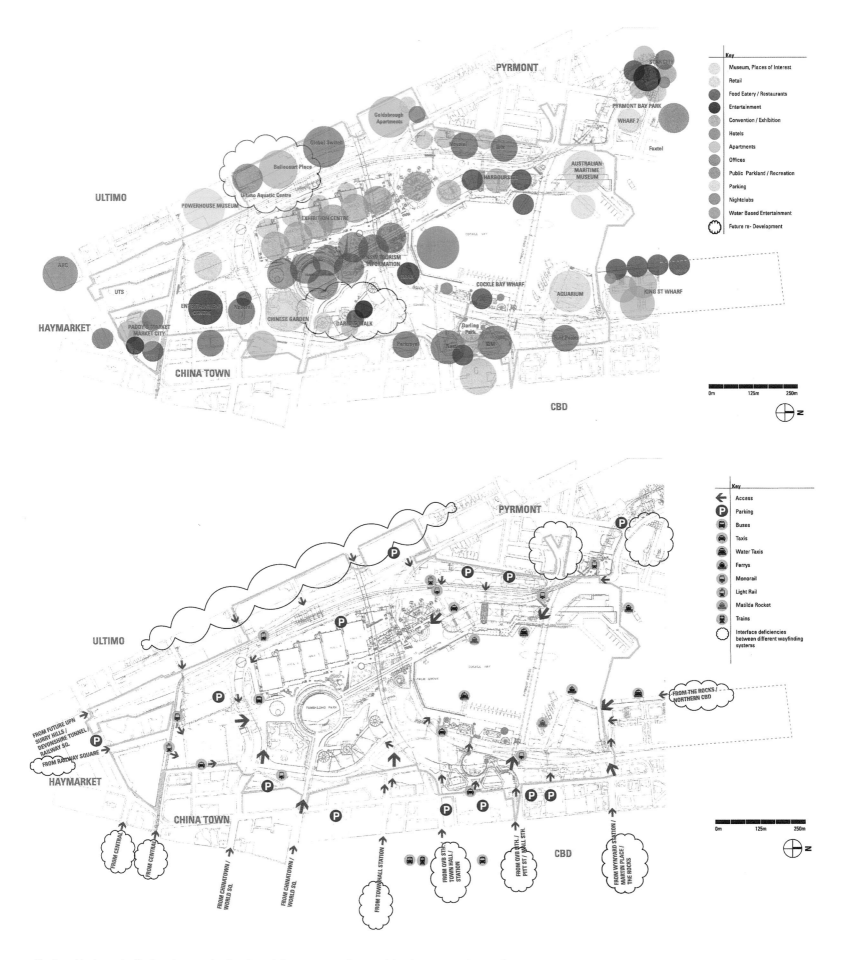

Darling Harbour is Sydney's premier tourism, leisure, recreation and business events precinct covering 60 hectares and attracting more than 25 million people movements annually. Darling Harbour commissioned a review of the signage taking into account the considerable developments and changes in and around Darling Harbour that have occurred over the previous 12 years. This resulted in additional and improved layers of signage ranging from identification to information and direction. Additionally a map of the precinct was prepared.

达令港是悉尼最大的旅游、休闲、娱乐和商业项目，占地60,0000平方米，每年吸引了超过25万人的客流量。

达令港委托设计师对达令港标识重新规划，在过去的12年里，其自身和周围已经发生了改变。这额外促使了标志的改进，范围从识别到信息和方向。此外，这个地区的地图也准备就绪了。

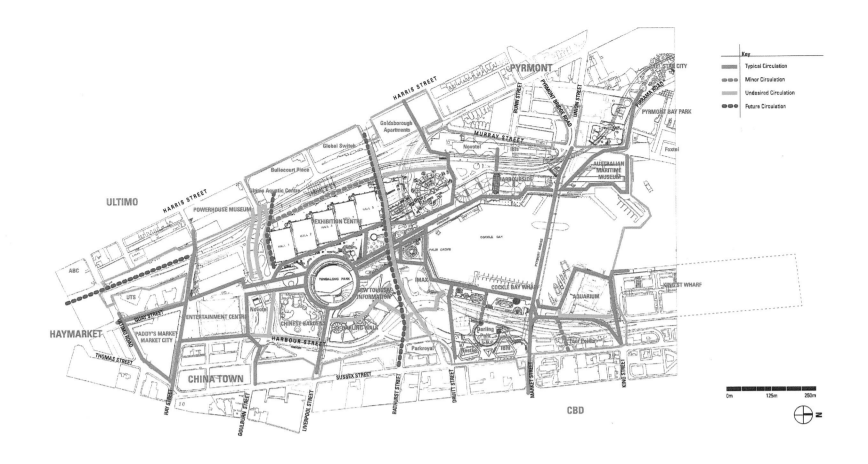

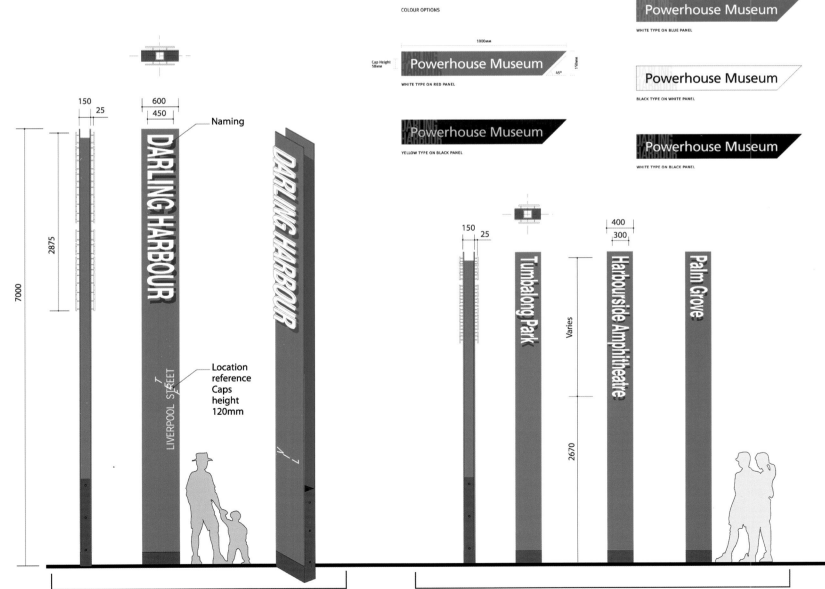

COLOUR OPTIONS

Powerhouse Museum
WHITE TYPE ON RED PANEL

Powerhouse Museum
YELLOW TYPE ON BLACK PANEL

Powerhouse Museum
WHITE TYPE ON BLUE PANEL

Powerhouse Museum
BLACK TYPE ON WHITE PANEL

Powerhouse Museum
WHITE TYPE ON BLACK PANEL

Naming

Location reference Caps height 120mm

DARLING HARBOUR

LIVERPOOL STREET

Tumbalong Park

Harbourside Amphitheatre

Palm Grove

Entry Marker

Place Identification

DESIGN COMPANY	LOCATION	AREA
Global, Arquitectura Paisagista lda.	Lisbon, Portugal	63, 000m²

Bicycle Path

自行车道

The project is part of a vaster strategy of a cycling network which is being developed by the municipality of Lisbon. However, the relation with the city's harbour and the river Tagus sets forth a unique identity for this particular route. The heterogeneity of spaces and environments, the proximity or crossing of several industrial and monumental sites drive the designers to approach this place with caution and sobriety and to determine exactly what the space through which this route wanders about is.

There are conficts because of the presence and characteristics of so many different surfaces associated with so many contexts which, on the other hand, have an important role in the construction of city's image and memory and its relation with the river. This acknowledgment took the designers to pursue a careful study of the successive layers and covers which were accumulated on the river bank in order to find a common ground and to produce a clearer and intenser image.

The design lays down a reversible bicycle trail over the memory of the river bank, the city and the river itself in order to minimise conficts with other mobility modes and to imprint an unmistakable route.

The strategy first lined out by the municipality is accepted totally and with rigour. In addiction the project interprets the clarity of its linear form with a system of signs, impressions and incisions on the vast mosaic of pre-existed or introduced surfaces. Communication, movement and experience become space attributes for the riparian city borders.

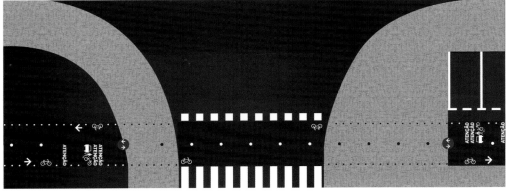

PE_08_A
esc. 1/100

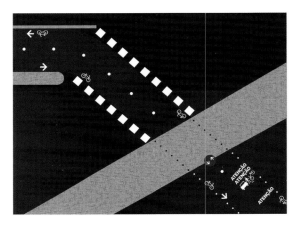

PE_08_B/C
esc. 1/100

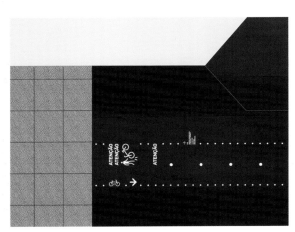

DIR-10
esc. 1/100

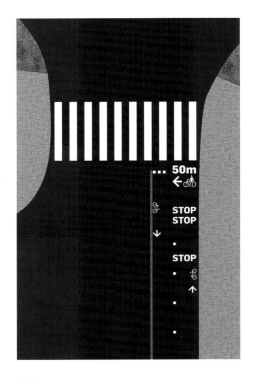

PE_09_A
esc. 1/100

PE_09_B
esc. 1/100

PE-01_B
esc. 1/100

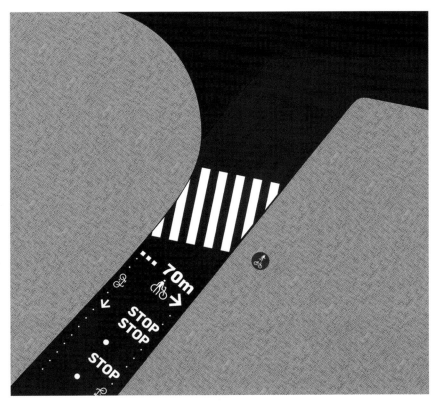

PE-01_C
esc. 1/100

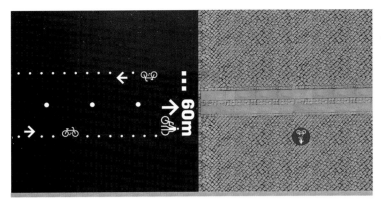

PE-02_A
esc. 1/100

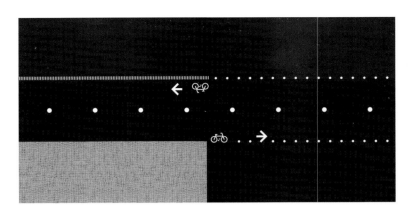

PE_02_C
esc. 1/100

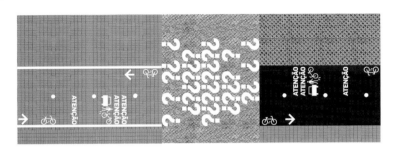

PE-04_A
esc. 1/100

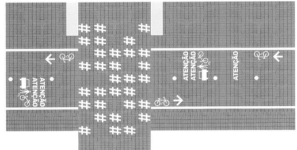

PE-04_B
esc. 1/100

PAREDE 01
esc. 1/100

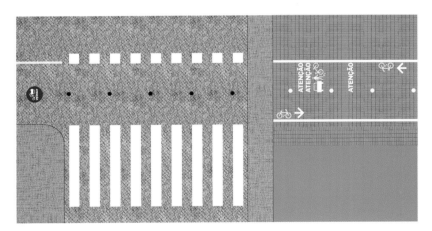

PE-04_C
esc. 1/100

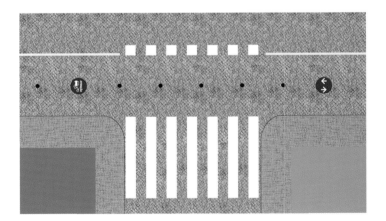

PE-04_D
esc. 1/100

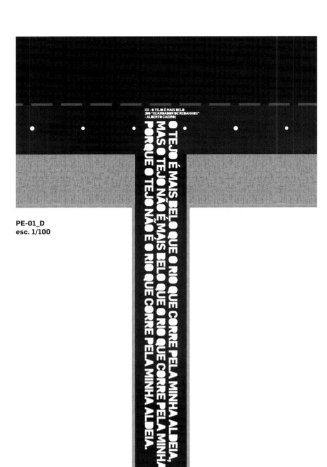

PE-01_D
esc. 1/100

PE-01_A
esc. 1/100

PE-09_D
esc. 1/100

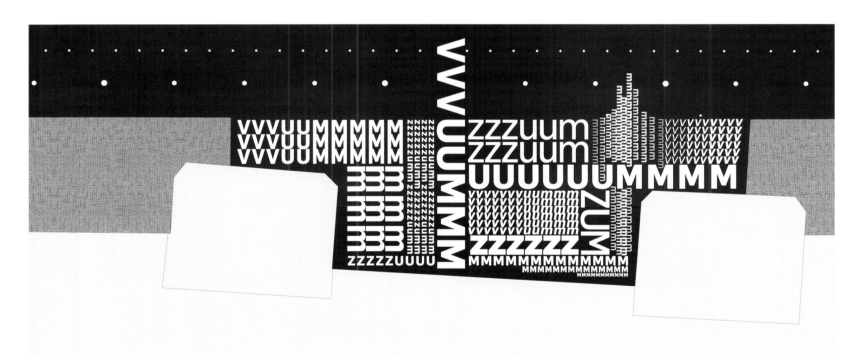

PE-05_C
esc. 1/100

PE-08_D
esc. 1/100

PE-02_B
esc. 1/100

O TEJO DESCE DE ESPANHA

E O TEJO ENTRA NO MAR EM PORTUGAL.

TODA A GENTE SABE ISSO.

MAS POUCOS SABEM QUAL É O RIO DA MINHA ALDEIA

E PARA ONDE ELE VAI

E DONDE ELE VEM.

E POR ISSO PORQUE PERTENCE A MENOS GENTE,

É MAIS LIVRE E MAIOR O RIO DA MINHA ALDEIA.

O TEJO TEM GRANDES NAVIOS. E NAVEGA NELE AINDA, PARA AQUELES QUE VÊEM EM TUDO O QUE LÁ NÃO ESTÁ, A MEMÓRIA DAS NAUS.

该项目是里斯本市当局开发的自行车网络更广大战略的一部分。然而，城市港口与塔霍河的关系为这条特定的道路向外界阐述了一种独特的标识。场地和环境的差异性，一些工业场地和努比亚遗址的附近或交界处引领设计师进入这个谨慎而严肃的地方，并准确定通过这条道路会漫游到什么地方。

正因有多种环境造就的如此多不同特征的地貌，所以才有冲突。另一方面，它在城市的形象和记忆、与河流的关系建设中起着重要作用。这样的认知让设计师追寻一个对累积在河岸上的连接层与覆盖物的仔细的研究，以便找到一个共同层面，喷出一个更加清晰和更加鲜艳的图像。

该设计在河畔、城市与河流自身的记忆中铺设了一条双向的自行车道，为了尽量减少与其他机动车辆的冲突，工人们压印了一条明确的路线。

这个战略首先被市政府划出，并经过谨慎的考虑后完全通过。此外，该项目用符号系统、压印和在已有的或新引入的地面上巨大的马赛克雕刻，清晰地诠释了它的线性形式。交流、运动和体验成为了河岸城市边缘的场所属性。

ELEMENTOS SINALÉTICA / SINALÉTICA DIRECCIONAL INCRUSTADA
esc. 1/10

CORDOARIA NACIONAL

0 KM
7 KM

ATENÇÃO

COMPOSIÇÕES DE SINALÉTICA PERIGO
esc. 1/100

ATENÇÃO
ATENÇÃO

ATENÇÃO

ATENÇÃO
ATENÇÃO

ATENÇÃO

... 35m

ATENÇÃO
ATENÇÃO

ATENÇÃO

Intelligent Traffic Lights

智能交通信号灯

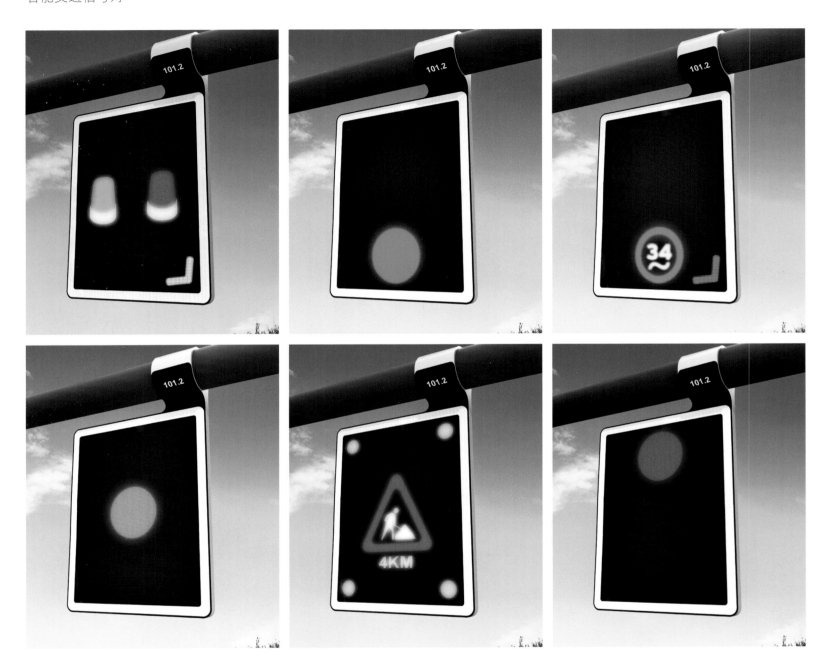

Traffic lights have been the same for over thirty years. Dennis van Melick believed it was time for update!

He designed a completely new traffic regulation system based on Red Green Blue LEDs, which guides road users through traffic more safely than ever. "Three to One: Traffic display is more compact, more efficient, and boasts more features than current traffic lights," explains Van Melick. Red is a square, orange a triangle and green a circle: colour-blindness is no longer a problem. With its clear signals the display can close a lane, create a bus lane, or signal ambulances, police cars and fire engines. An innovative suspension system makes installation easy and vandal-pro.

在过去的三十年里，交通信号灯从未有过改变。丹尼斯·范·梅里克认为是时候要更新换代了！

他以红绿蓝 LED 灯为基础，设计了一款与以往相比安全性更高的新型交通管制系统，指挥行人来往。"三合一交通指示器更加简洁、高效，比通用的交通信号灯功能更多，" 范·梅里克解释说。红灯是正方形，黄灯是三角形，绿灯则是圆形，这样一来即使是色盲也能看懂。指示器上信号灯清楚明确，在此情况下可以关闭一个车道，建立一个公交车专用车道、或者利用信号通知救护车、警车以及消防车。这种悬架系统不仅具有创新性、易于安装，还能免去遭人为破坏的烦恼。

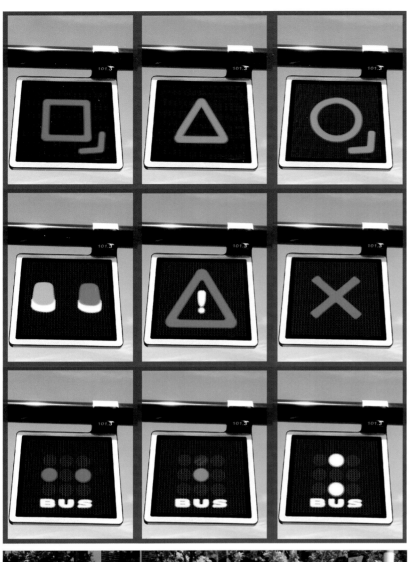

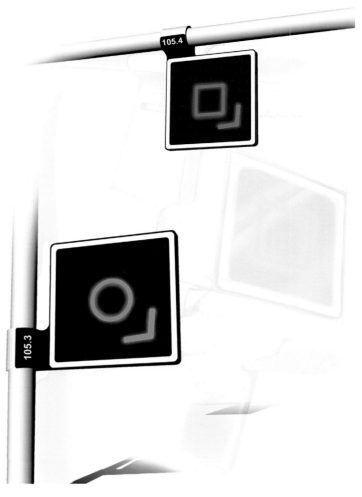

Info Display

信息显示牌

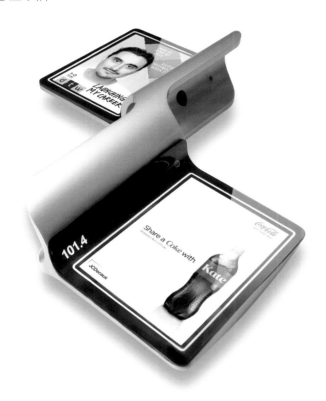

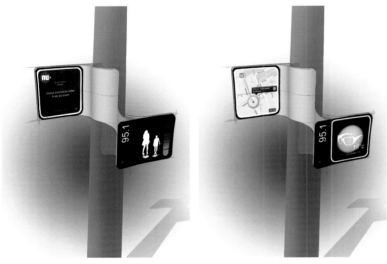

Digital Information System:

This Info Display system is part of the new traffic regulation system designed by Dennis van Melick. It is a multifunctional digital display that provides pedestrians and cyclists with useful information on the road. The system not only replaces the button to push at the lights, but also gives you weather and news reports while you are waiting for the lights to turn green. The stand-alone version does well at public transportation stops too; while waiting for your tram or bus, this will show you the cultural calendar of the city. Messages change automatically, so an endless stream of information will be included in the loop.

数字信息系统：

该信息显示牌是丹尼斯·范·梅里克设计的交通管制系统的一部分。它是多功能的数字显示，为行人和骑自行车的人提供有用的信息。当你正在等待信号灯变绿时，该系统不仅取代了红绿灯的按钮，还给你带来天气和新闻报道。公共交通站点的独立系统同样应用得很好，当你在等电车或巴士时，它会告诉你所在城市的文化历史。信息不断地自动更新，这样不间断的信息源显示在信息牌上。

DESIGNER	DESIGN COMPANY	CLIENT	LOCATION	PHOTOGRAPHER
Hans Gerber and Yasmin Hall	Minale Tattersfield	Attorney General Department	Sydney, NSW, Australia	Minale Tattersfield

Parramatta Justice Precinct

帕拉马塔司法管辖区

The Parramatta Justice Precinct provides community access to a wide range of metropolitan-based justice services. Effective signage is extremely important in assisting the public to locate the various services.

An external signage strategy was developed to provide an integrated family of signs addressing the needs for orientation, building and precinct identification.

A Braille & tactile map was prepared for the precinct.

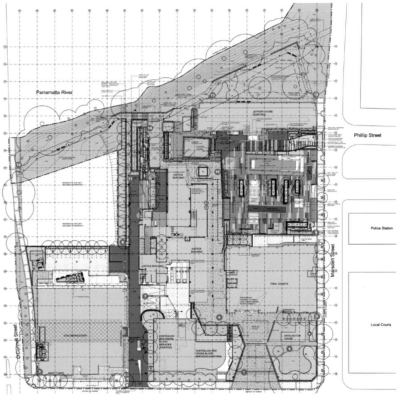

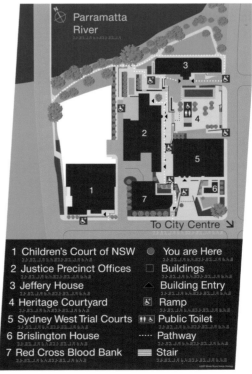

帕拉马塔司法管辖区规划了大范围的城市基础司法服务。有效的指示标识因此尤为重要，可以帮助市民确定不同服务的地理方位。

外部的标识方案为城市提供了一个完整的标识系统，可以来帮助市民确定方向、定位建筑、以及熟悉不同服务设施边界。

盲文和触觉地图在管辖区内也有设置。

DESIGNER

Ruedi Baur

DESIGN TEAM

Intégral Ruedi Baur Paris – Ruedi Baur, Stephanie Brabant, Eva Kubinyi, Claudia Leuchs, David Thoumazeau

Metz Signage

梅斯市标志

To mark the opening of the Metz Pompidou Centre in 2010, Ruedi Baur and staff produced temporary signage linking the museum with the city centre. Further to this initial project, the Metz City Council then entrusted Intégral Ruedi Baur with designing the signage for the entire city.

Starting with a concept of messages written in space, the typographic composition is laid on horizontal lines that both link the letters and hold them up. The whole can stand in its own right, adorn an existing feature or actually mark a building. The Irma font appears to thread over the stretched clear frames, in order not to conceal the city's pre-existing architecture.

Unlike the temporary signage, which was decked out in pastel hues, this permanent signage is white, enhanced with touches of bright and even fluorescent tints. The single or double posts are white, with colour painted on the edges and inside the lettering, the colour being chosen to fit the environment.

Each directional sign is water-jet cut in a single piece from an aluminium block, meaning that there are no soldering marks or joins to mar the beautiful precision in the use of colour painted within each letter and on the edge of each word.

There are eight separate tourist circuits, starting at the railway station and fanning out towards the city centre, neighbourhoods and the Metz Pompidou Centre. As from 2013, there will be signs denoting the entrance to pedestrian districts, signage to identify the entrances to covered markets, like typographic canopies, or over-hanging lines announcing the various neighbourhoods.

The concept of this new informational and directional signage system is to depict the city as it is, slipping into the context, revealing it and endeavouring to promote it.

A poetic approach for a fresh twist on the city of Metz and its riches.

CLIENT

Metz City Council

RESEARCH AND DRAFTING OF COPY

Thibault Fourrier

LOCATION

Metz, France

2010 年，为表明梅斯市蓬皮杜中心正式对公众开放，设计师 Ruedi Baur 及其团队制造了临时标志，将博物馆与市中心连接起来。继这一最初的项目之后，梅斯市议会委托 Intégral Ruedi Baur 设计团队为整个城市设计标志。

基于太空中的留言板理念，该排版组合被放置在水平线上，不但将各个字母连接起来，而且将它们固定住。标志整体可以独立站立，装饰着现有的外观，或者说实际上它是一栋大楼的标志。伊尔玛字体似乎连接起镂空的清晰边框，而不会遮挡住城市先前的建筑。

与用柔和色调装扮的临时标志不同，永久标志呈白色，用高光甚至荧光色调进行加强。单杆或双杆是白色的，其边缘与字母内部均涂有颜色，所选的颜色与周围的环境协调呼应。

每个方向标志都采用水射流切割工艺，从铝块上直接切下来。这意味着在每个字母内部及每个单词边缘都不会出现焊接痕迹或接合点，因此涂在上面的彩色油漆精准完美。

这里共有八条独立的旅游环路，从火车站开始，朝着市中心、居民区及梅斯市蓬皮杜中心，呈扇形散开。自 2013 年起将会有标志指示步行区与室内市场的入口方向、有的像印刷排版体的罩幕，有的像向外伸出的直线，标志着不同的街道。

这一新型的标志系统可以提供信息、指明方向。其理念是要如实地描写城市、与环境融为一体，展现城市的面貌，并争取使其不断发展。

这一富有文采的创举对于梅斯市及其房地产来说都是一次新的转折点。

PROJECT DESIGNER
Diana Cabeza, Martín Wolfson, Leandro Heine

GRAPHIC DESIGNER
Osvaldo Amelio Ortiz, Gabriela Falgione, Pablo Cosgaya, Marcela Romero

Street Signage System

街道标识系统

Urban elements must respond to the geographical and cultural environments that generate them and be able to blend with its general disposition and particularities.

Conceived as an integral and inclusive system, this signage system is intended to be placed on all the city of Buenos Aires as a whole. Designed with the idea of flexibility, it can be installed attached to the walls of building façades, on sidewalks on a self-standing support, or fixed to the traffic light structures. This flexible approach creates continuity and legibility in the context of an existing city with significant constraints and much visual pollution.

The project focuses on preserving the historical heritage of the city, while providing modern urban elements for everyday life. In the case of self-standing signage, they are designed addressing a classical order, recovering the traditional technology of urban elements, whilst updating industrial fabrication and maintenance procedures.

The whole system of urban elements was created with the idea of accessibility. The street signage system with the bus shelter system (see page 36) constitutes together the whole urban furniture for the city of Buenos Aires.

CLIENT
Government of the city of Buenos Aires.

MATERIAL
Cast Iron, painted steel structure, aluminum extruded sections painted black with reflective letters

LOCATION
Buenos Aires, Argentina

城市元素应当与孕育它的地理与文化环境相呼应，并且能够融入环境的整体布局和特质个性。

本标识系统的最初构想是建成一个覆盖整个布宜诺斯艾利斯市的完整的、包罗万象的系统。

标识系统的设计具有灵活性，可以附着安装在建筑的表面，通过自动支架立在人行道旁、或者固定在交通信号灯上。

布宜诺斯艾利斯市现存着很多的制约因素和视觉污染，但是正是因为有着灵活的设计，使得标识系统在这样的条件下仍然具有连续性和可识别性。

这项工程在发展日常生活所需的现代城市元素的同时，保存该城市的历史遗产。

不依附于建筑的独立标识被设计安装在古典柱上。它们既重现了传统城市工业技术，又体现出全新的工业制造和维修技术。

整个城市元素系统都建立在环境可达性的理念之上。

标识系统和公交候车亭系统共同构成了布宜诺斯艾利斯市的整套地景设施。

DESIGNER
Ben Busche

TEAM
Joao Gómez Leitao, Antonio Bernacchi, Giovanni Maria Biddau

CLIENT
City Council Madrid

CONSTRUCTOR
Disval SA

Sign System de.dos

de.dos 标识系统

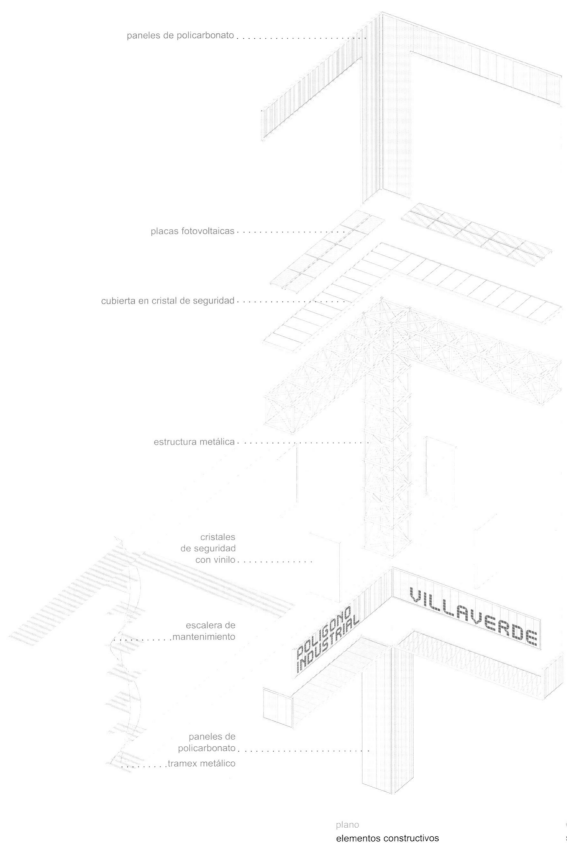

paneles de policarbonato . . . .

placas fotovoltaicas . . . .

cubierta en cristal de seguridad . . . .

estructura metálica . . . .

cristales
de seguridad
con vinilo . . . .

escalera de
. . . . .mantenimiento

paneles de
policarbonato . . . .
. . . . .tramex metálico

VILLAVERDE

POLIGONO
INDUSTRIAL

plano
elementos constructivos

escala
s.e.

LOCATION
Madrid, Spain

DIMENSION
2m x 2m (base) x 16 m (height)

PHOTOGRAPHER
Miguel de Guzman, Brut Deluxe

The sign system for industrial parks totem de/dos has been conceived in order to guide visitors who are not familiar with the site visually to their destiny in clear and evident way.

The structure is composed of galvanised metal profiles. The cladding is entirely of white coloured 40mm thick cellular polycarbonate plates fixed on the perimeter with white coloured aluminium profiles corresponding to the respective panel system. The exterior plane of the primary structure and the interior plane of the polycarbonate cladding are separated 60mm in order to locate LED stripes in the interstitial space and achieve a homogenous illumination on the interior. The two inferior modules of all posts is realised in security glass to offer a greater resistance against vandalism. Printed vinyls with the graphic information are fixed on the inner side of these glassed modules. The roof is realised also in horizontally laid security glass fixed punctually on the secondary structure. In the interior of the structure a series of stairways are located connecting different levels of galvanised metallic grids that give access to the installations and its maintenance. Both posts are illuminated in its interior by white coloured LED stripes. The illumination is activated simultaneously with the street lighting of the industrial district. A distinctive colour of the lightning for each industrial park was proposed originally though finally all will be illuminated in white. On top of the primary posts has been foreseen the optional installation of photovoltaic panels without having been realised so far. All posts are equipped with graphic information – a map of the district and a business listing.

标识系统是为 totem de/dos 工业公园设计的，为引导不熟悉那里的游客，它以清晰、明确的方式在视觉上指引他们前往目的地。

它用镀锌金属型材构成。包层完全是白色的 40 毫米厚的聚碳酸酯板，固定在白色的彩色铝型材上，对应于相应的面板系统。原先结构的外平面以及聚碳酸酯熔覆的内平面被分离 60 毫米，以确保条纹有足够的间隙，在内部实现均匀的照明。全部标牌下方的模块用安全玻璃的覆盖来提供更好的抗破坏性。有图形信息打印的乙烯基固定在这些玻璃模块的内侧。顶部的搭建以水平放置安全玻璃制成精准设置的二级结构实现。在建筑内部一系列的楼梯被设置来连接镀锌金属网格的不同层次，以便提供对建筑的设置和维护。两个部分都在内部用白色的 LED 灯条点亮。照明运行的同时，工业地区的路灯也相应点亮。每个工业公园独特颜色的照明方案最初提出，最终所有的工业区被笼罩在白光中。

在最初建筑的顶部已经预见了安装太阳能板的可能性，所以并没有过多设计。所有的部分都配以图形信息——一个街区和公司陈列的地图。

DESIGNER
Ignacio Ciocchini, IDSA. Industrial Designer

MANUFACTURER
Temple Inc.

LOCATION
Chelsea, New York , USA

Streetsigns

街头标志牌

Problem: Standard New York City streetsigns are too small and difficult to read at a distance particularly after dark.

Solution: Advantages of the new Self-illuminated Streetsigns

• The blade size was doubled, from 38" X 7 ½" to 50" X 14"

• Easily readable upper and lower case Frutiger 65 font can be seen from a block away

• The size of the upper case font was increased from 4" to 6 ¼"

• LEDs inside the extruded aluminium frame illuminate the sign board at night with a soft glow

• Energy consumption: $9 a year per sign. 1KW (100 Hours)

• Block numbers with arrows were included at the top of the sign to provide useful wayfinding information

问题：标准的纽约街头标志牌面积过小，尤其在黑夜远距离很难识别。

解决途径：新的照明街头标志的自身优势

• 标志平面的大小增加了一倍，从 38"X 7½" 变为 50"X 14"

• 易读的大写和小写字体 Frutiger 65 可以从一个街区外识别

• 大写字体的大小从 4" 增大至 6¼"。

• 发光二极管设置在铝框架内，在夜间用柔和的光辉照亮标志牌

• 能源消耗：每个标志牌每年 9 美元。1kW（100 小时）

• 带箭头的街区号码设置在标志牌最上面，提供寻找道路的有用信息

MATERIAL Powder-coated extruded aluminium frame, polycarbonate panels, vinyl graphics, and LEDs

DIMENSIONS
50" x 14" blade, hump height is 20"

PHOTOGRAPHER
Marco Castro

DESIGNER
Hans Gerber

DESIGN COMPANY
Minale Tattersfield

CREATIVE DIRECTOR
Hans Gerber

The Oracle, Broadbeach

The Oracle 地标建筑

Branded signage for one of the world's most sophisticated ocean front beachside precincts at the Gold Coast, Australia's most dynamic resort city.

The Oracle Precinct occupies about 13,000sqm and is structured into four unequal 3-level podium buildings by an internal boulevard, a lane and a square. The two larger building blocks feature apartment towers with 36 and 46 levels. Each block has an entry lobby for corporate tenants. The majority of ground floor street frontage is made up of 44 boutiques, cafés and restaurants.

CLIENT
Niecon

LOCATION
Broadbeach, QLD, Australia

PHOTOGRAPHER
Minale Tattersfield

品牌标识是为澳大利亚黄金海岸上一个世界上最复杂的海景海滨区设计的，它也是澳大利亚最具活力的旅游城市。

Oracle 建筑区占地约 13,000 平方米，通过一个内部林荫大道、一个车道和一个广场，区域建造成拥有四个在三个水平高度的裙楼。

两个较大的裙楼分别设有 36 层和 46 层公寓。每栋大楼有一个专为公司租户设计的入口大厅。一楼临街的绝大部分是由精品店、咖啡馆、餐馆等 44 个店铺组成。

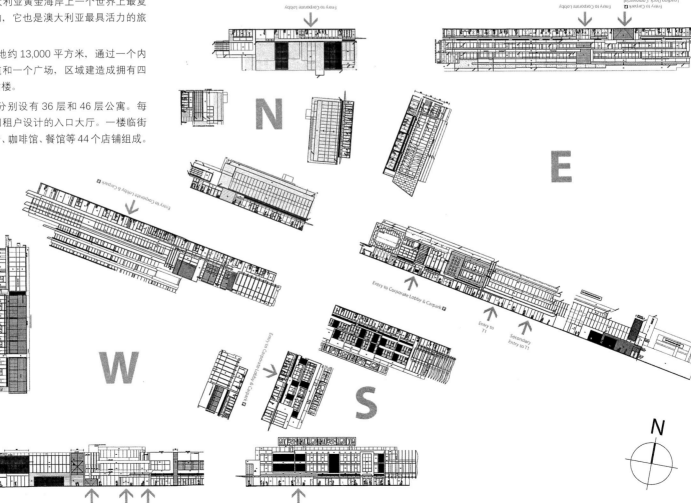

DESIGNER
Hans Gerber, Jessica Tse

DESIGN COMPANY
Minale Tattersfield

CREATIVE DIRECTOR
Hans Gerber

Randwick Civic Markers

兰德威克城市标识

Randwick is in metropolitan Sydney. It features a major university and hospital, large sporting complexes, some of Sydney's best-known beaches and Australia's largest commercial shipping port.

The challenge was to develop a design, which was cognisant of the significance of the city, its heritage, the present and future of the city. The design had to take the shape of an upright structure in order to be an optical holding point, to fit into tight spaces and to have the graphics arranged vertically, in order to be recognised from a distance. The design comprises of twin blades to enhance transparency and to structure the written information to depict the name of the city on one blade and the name of the suburb on the other blade. The application of the city's logo is applied as a watermark. The watermark is halved to fit equally to the blades. All text is in the highly legible Frutiger typeface in dark grey on silver background.

The markers are designed to express the duality of old and new, city and suburb, subject and object. The juxtaposition of the blades and the open vertical space between adds the notion of transparency. The signage programme also comprises park identifiers, featuring maps and regulatory signage.

兰德威克位于悉尼的市中心。它有一个重要的大学、医院、大型的运动配套设施、一些悉尼的著名海滩、澳大利亚最大的商业航运港。

挑战是进行一项设计，可以来表达城市的文化、当地的传统风俗以及城市的现在和未来的重要性。设计采用直立结构来在视觉上引人注目，可用于狭窄空间，有垂直排列的图形，这样，在很远的地方就可以识别出来。为提高透明度，设计采用两个薄片构建，在一面显示了城市名称的文字信息，另一面是周围郊区名称的信息。城市标识采用水印图案。图案分成两半来同等地适用于两个平面。在银色标识板上所有深灰色的字体采用高清度的Frutiger 字体。

标识设计是表达新与旧、主体与客体、城市与郊区的二元性。并行排列的薄片以及之间的垂直空间都提升了标识的透明性。显示的信息也包括停车标识、特殊地图、管理条例。

CLIENT
City of Randwick

LOCATION
Sydney, NSW, Australia

PHOTOGRAPHER
Minale Tattersfield

PUBLIC ART

公共艺术

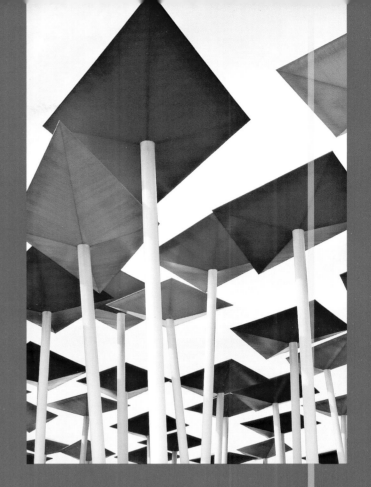

DESIGNER
Michel Rojkind, Gerardo Salinas

DESIGN COMPANY
rojkind arquitectos

TEAM
Arie Willem de Jongh, Alfredo Hernández

Portal of Awareness

咖啡杯通道

This portal, conceived through spatial design, activated by the city dwellers and the everyday stimuli of Mexico City's life, becomes a new public piece in one of the city's most important avenues, Paseo de la Reforma.

Nescafé commissioned eight artists (Rojkind Arquitectos, Francisco Serrano, Mario Schjetnan, Bernardo Gómez-Pimienta, Fernanda Canales, Manuel Cervantes, Alejandro Quintanilla and Alejandro Castro) to develop site-specific installations on Paseo de la Reforma, one of the city's most important avenues, with the basic requirement of utilising a maximum of 1,500 metal coffee mugs.

Combining a common everyday life object, a mug, with a basic common architectural construction material, rebar, the idea of a portal takes place. The rebar is used as the primary structure for the cups, which are then mechanically attached to each intersection of the steel mesh. Combining 41 main arches, ranging in lenght from 10 to 12 metres, with two additional layers of 56 diagonals each, the piece is inter woven to create 1,497 nodes for the cups.

The final shape of the portal, along with the different colours of the mugs selected, reinforce the sense of movement of the piece, which plays a key role in the concept of the project. The steel planters anchor the structure and allow for the vines to grow in between the rebar, with the idea that in time it will cover the entire structure in a green foliage from the outside, while the inside displays the gradient of the mug's chromatics.

The play of the shadow's patterns casted on the sidewalk add an extra layer that shifts throughout the day.

This installation is intended to be in place during the winter months, providing a space of expression and interaction in the public realm.

CLIENT
Nescafé

LOCATION
México City, México

AREA
42m²

PHOTOGRAPHER
Jaime Navarro

BUILDING SEQUENCE

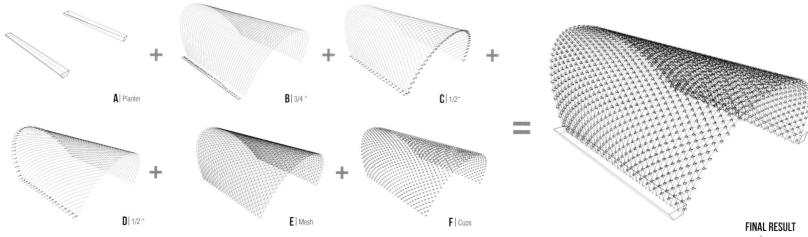

A | Planter   +   B | 3/4 "   +   C | 1/2 "   +

D | 1/2 "   +   E | Mesh   +   F | Cups   =   **FINAL RESULT**

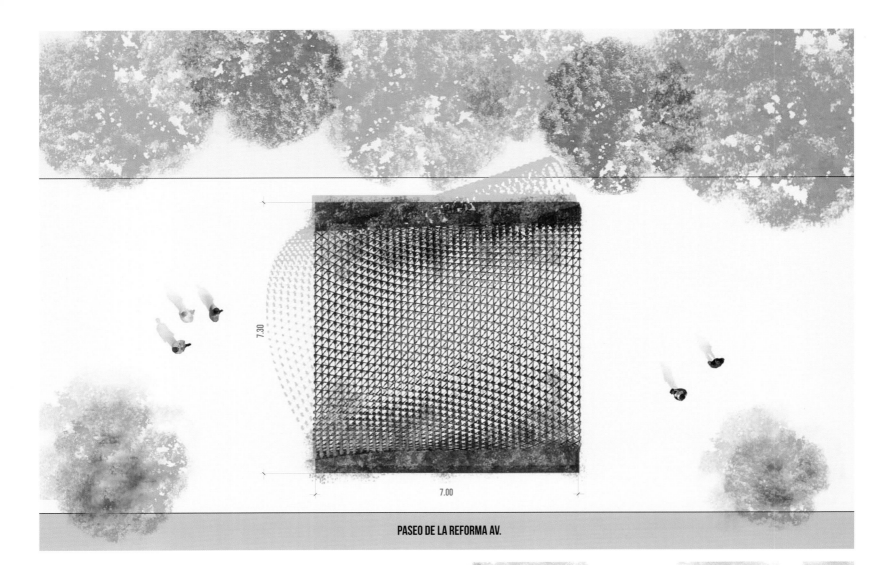

PASEO DE LA REFORMA AV.

这处通道采用立体构造，人来人往十分热闹。它为每天的墨西哥城市生活增添了一道亮丽的风景线，也逐渐成为城市最重要的街道 Paseo de la Reforma 上的一处新的公共地带。

雀巢公司委托八位艺术家（Rojkind Arquitectos、Francisco Serrano、Mario Schjetnan、Bernardo Gómez-Pimienta、Fernanda Canales、Manuel Cervantes、Alejandro Quintanilla 以及 Alejandro Castro），用 1500 个金属材质咖啡杯为墨西哥市最重要的大道——改革大道，设计制作了该装置。

把众多元素结合在一起：一项日常的生活设施、马克杯、钢筋——一种常见的建材，通道的想法就产生了。钢筋被用来构建主体框架，马克杯被整齐地粘贴在金属网格的每一个节点上。41 个长度 10~12 米的主要拱条，另外叠加两个分别拥有 56 条对角线拱条的拱层，共同产生了 1497 个节点来设置杯子。选取不同颜色的马克杯，通道的最终外观加强了设计的运动感，这也是这次项目的主要概念。金属花架既固定了结构，又可使藤蔓植物生长在钢筋之间，当藤蔓植物的绿色枝叶遍布外部框架，通道内部展示出马克杯色彩的变化梯度。

通道阴影被映在人行道上，增添了新的层次，阴影的映射图案在一天之中不断变化。

通道的设计是希望在冬天里为人们提供一处活动场所，让他们在公共空间里互动交流。

DESIGN COMPANY	CLIENT	DIMENSION	LOCATION	AREA	PHOTOGRAPHER
SLOT	Promexico	L 80m X W 50m	Shanghai, China	3,500m²	SLOT

Mexican Pavilion

墨西哥馆

The pavilion's design is born from the idea of representing Mexico through its traditional elements that haven't been exploited in these kinds of fairs. The proposal scheme is centred around the idea of creating a green space within the EXPO which at the same time represents SLOT's preoccupation to offer a better life standard for cities through the recovery of green areas rather than creating a protagonist building.

The Mexican Pavilion is a volume defined by a talud (slope) that transforms itself into a plaza privileging public space as an urban gesture within the EXPO. Space is divided in three levels that represent three different moments of urban life in the country. The past is represented on the plinth, present time Mexico at the entrance level, and future on the platform.

The pavilion's main feature lies within the design of the papalotes (kites), a word that comes from the Nahuatl "papalotl" that means butterfly, used as a cultural meeting point between mexican and chinese cultures.

The proposal is to look into a future with areas that are thought, destined and planned specifically for leisure, the recovery of parks and green areas, where new generations might meet in a city with a "better living".

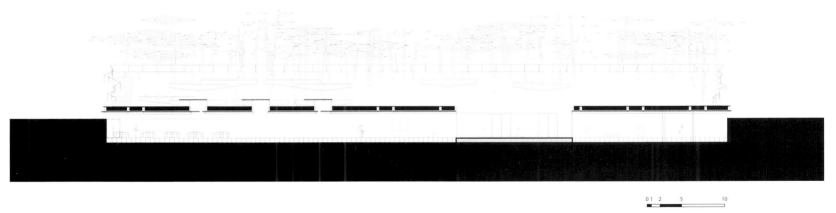

0 1 2   5   10

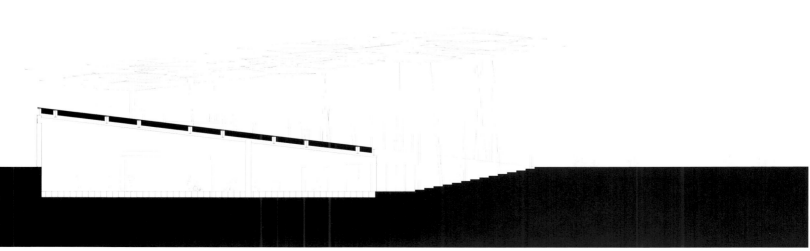

0 1 2   5   10

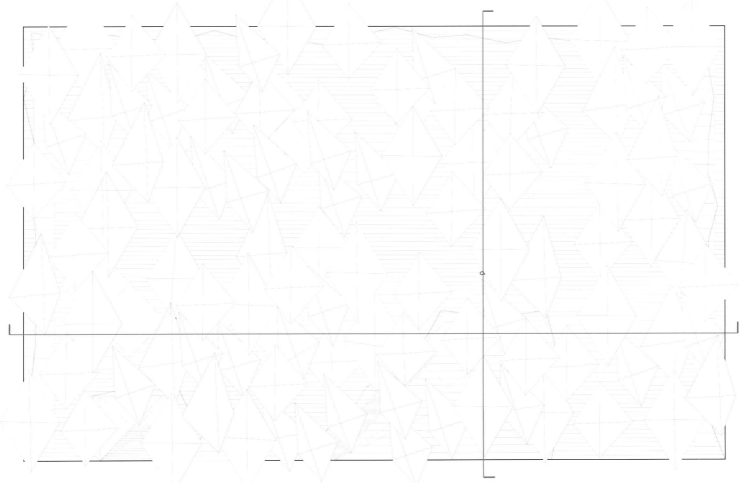

01 2   5   10

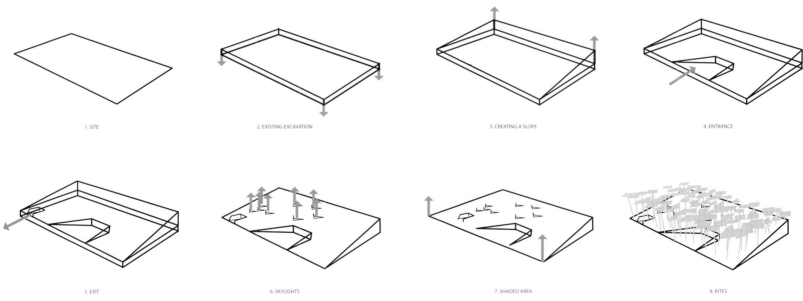

1. SITE   2. EXISTING EXCAVATION   3. CREATING A SLOPE   4. ENTRANCE

5. EXIT   6. SKYLIGHTS   7. SHADED AREA   8. KITES

通过传统元素来展现墨西哥，是该展馆设计的灵感来源，而这些传统元素之前是没有在此类展馆采用的。提议的方案是围绕世博会创建一个绿色空间的理念，同时建筑要能表现对我们提升城市生活品质的深切关注。我们对城市生活品质的提升要通过还原绿地来实现，而不是建造更多的高楼大厦。

墨西哥馆是一个以 talud（斜坡）为特征的建筑，在世博会上它作为城市标志，变成一个类似广场的公共场所。展馆空间分为三层，代表了墨西哥城市生活的三个不同时期。底层展现了城市生活的过去，与通道水平的中层代表了现在的墨西哥，最高一层的平台展示了墨西哥城市生活的未来。

展馆的主要特色体现在 papalotes（风筝）景观的设计上，papalotes 源自纳瓦特尔语，代表蝴蝶的意思，它被作为一个中国文化和墨西哥文化交汇的地点。

展会延伸至未来，很多区域被特地安排规划为休闲场地、公园以及绿地，青年一代也许就可以在未来的城市中看到，从此过上"更加幸福的生活"。

 **ARTIST**   **LOCATION**

Thilo Frank   Hjallerup, Denmark

Ekko

Ekko 设施

The latest work of Berlin-based artist Thilo Frank invites the visitor to explore his spatial perception through movement and sound.

Ekko consists of 200 wooden frames revolving once around their own axes around a circular concrete path. This inner form is enclosed by a fence like structure consisting of another 200 wooden poles. Depending on the daylight the shadow play creates alternating patterns – from further distance the sculpture flickers in a moiré effect. The visitor is challenged to perceive and explore the three-dimensional shape.

While moving through the sculpture – built-in microphones pick up the sound of the visitors and a computer sound system filters and remixes the recorded sound and sends it to built-in electrodynamic resonators – the sculpture becomes an instrument and plays a constantly changing soundtrack. The work acts as an archive of sounds and at the same time the visitors' perception of space and presence is amplified.

柏林的基层艺术家 Thilo Frank，邀请参观者们通过动作和声音来探索他最新作品的空间知觉。Ekko 设施是由 200 个木质框架搭建成的，木质框架在圆形的混凝土小道上绕自己的轴线旋转了一圈。这种内在的形式是封闭栅栏状结构组成的，包括另外 200 根木棍。根据日光的不同，光影循环交替——从更远的距离看，这个作品闪烁着莫尔效应。参观者面临的挑战是感知和探索三维空间。

当在这个作品里穿梭的时候——内置麦克风就会采集参观者的声音，电脑声音系统会过滤并重新合成这些声音，然后发送到内置电动谐振器上——该作品就变为一个乐器，演奏不断变化的声道。它的作用就像声音储存器，与此同时，参观者对空间和现场的感知就得到了放大。

DESIGN COMPANY	CLIENT	LOCATION	AREA
ANNABAU Architecture and Landscape	Landesgartenschau Norderstedt Gmbh	Norderstedt, Germany	100 m²

Landesgartenschau Norderstedt-Religious Burial Traditions - Islam

诺德施泰特园艺——伊斯兰宗教葬礼的传统

The gardens, exhibiting burial traditions, are located on a former garden site with a beautiful tree population. A winding pathway is connecting these five different gardens.

This recurring exhibition entry of grave cultures, which is a part of every Landesgartenschau, has been re-interpreted showing gardens of all five world religions: Buddhism, Hinduism, Judaism, Islam and Christianity. Various religious traditions are presented in a condensed manner. These gardens are to be experienced as sculptural objects but also aim to raise interest in the different burial traditions.

In Islamic tradition, people are encouraged to spend their money on the poor and needy rather than expensive grave monuments. Graveyards in Islamic countries are therefore mostly calm and nature-oriented places which could be compared to a meadow.

The dead body is wrapped in linen and laid down on its right side looking towards Mekka. The ground above the grave is slightly culminated and the grave marked by a stone. Islamic graveyards are eternal resting places.

This Islamic tradition has been translated into a green and grassy garden. Silky waves of grass are enclosed by steel curbs. These waves are oriented towards Mekka. The surface of the synthetic resin-based gravel paths, cutting through the grass, is mimicking the movement of the ground which strengthens the orientation.

这些花园，建于一个有美丽的树群的花园旧址上，展示着这里的葬礼传统。蜿蜒的小道将这五个不同的花园连接起来。

这种循环式的坟墓文化的展览入口，是园艺展每个有机组成部分，它已重新诠释所有五种世界宗教——佛教、印度教、犹太教、伊斯兰教、基督教——的园艺展。以简明的方式呈现了各种宗教传统。

这些园林有如雕塑般丰富经验，也旨在提高不同的葬礼传统的趣味。

在伊斯兰传统中，宗教信仰鼓励人们将钱花在穷人和有需要的人身上，而不是花在昂贵的墓碑上。

因此，伊斯兰国家的墓地多为平静和纯自然的场所，可以比喻为草地。

尸体包裹在亚麻布里，朝右侧放置，面向麦加城的方向。坟墓的表面轻微隆起，石块是坟墓的标记。伊斯兰墓地是永恒的安息之地。

伊斯兰的传统墓地如今被建成长满绿草的花园。植草柔滑的坡被钢材板隔离着，坡面都朝向麦加城。合成树脂的表面铺垫了砾石小道，穿过草坪，模拟地面运动增强了方向感。

DESIGNER
Miguel Costa

DESIGN OFFICE
MAIpublicspace Research Platform

COLLABORATORS/ACTIONS BY
Meireles de Pinho, Artist; Filipe Garcia, Artist; and Crestuma's inhabitants

Open Monument

公共纪念碑

The installation of a "monument" that revisits the image of a maize crib on the riverfront attempts to restore an old relationship between some of Crestuma's existing maize cribs and the river Douro.

Further, this new element intends to become a landmark giving a new visibility to the town from the river, and, simultaneously, leaving its appropriation open by allowing room for collective actions from inhabitants and artists, who may intervene, modifying it continuously, proposing thus new authorships, new social dynamics and new senses of belonging – an "Open Monument".

The "Open Monument" project is a public space installation that intends to recover the link between the inhabitants of Crestuma and its landscape symbols/history. At the same time, it intends to boost an experimental, pedagogical and participative realm that could develop new senses of appropriation and belonging.

SPONSOR
Madep/Fbaup, Trans_Form_Actions European Project and Crestuma's local government

MATERIAL
Metal structure and perforated metal sheet

LOCATION
Crestuma Vila Nova de Gaia, Portugal

PHOTOGRAPHER
Miguel Costa

纪念碑是仿照竖立在河岸边玉米仓的样子建造的，希望可以重建 Crestuma 地区现存的一些玉米仓和杜罗河的的亲密关系。

此外，这个新建的构造希望能够成为一处地标，可以让河对岸城镇的人们清楚地看到。同时，它很大程度上对外公开，允许当地居民、艺术家前来对它进行修饰。逐渐地，人们将自己定义为纪念碑的拥有者，产生一种归属感，纪念碑更推动社区迸

发出新的活力。

"公共纪念碑"是一项公共空间设施，项目致力于恢复 Crestuma 地区的居民与该地区风貌／历史的联系。同时，希望可以起到现实性、教育性、群众参与性的作用，从而使人们建立起一种归属感。

DESIGNER
7N Architects and rankinfraser landscape architecture

COLLABORATOROR
Speirs and Major, Lighting

1 - Motorway carriageway overhead
2 - Proposed hotel
3 - Garsube Road
4 - Landscape Link
5 - Phoenix Flowers
6 - Terraces
7 - Proposed future link to canal
8 - Forth and Clyde Canal

The Phoenix Flowers, Garscube Landscape Link

凤凰花，Garscube link 景观设计

The Garscube Landscape Link is the first phase of the Speirs Lock regeneration initiative instigated by the Glasgow Canal Regeneration Partnership (GCRP), a partnership between Glasgow City Council and ISIS Waterside Regeneration, supported by British Waterways Scotland. The project brief called for the radical revitalisation of a crucial connection between the Glasgow's canal network and the city centre which had been severed by the construction of the M8 motorway in the 1960s.

The "Phoenix Flowers" project delivers this transformation of the public realm under the M8 at Garscube Road. This new link acts as the gateway point to Speirs Lock neighbourhood and the Forth and Clyde Canal, a pedestrian threshold connecting a large area of north Glasgow back to the city centre.

The existing underpass was an extraordinarily hostile environment: dark, noisy, dirty and intimidating. The proposals widen the space considerably, transforming it with a flowing, red resin-bonded surface that doesn't constrain those using it to a single, confrontational route. The surface also unifies the two sides of the space which were given different treatments according to their location. Prior to the construction of the motorway the site had been a park – Phoenix Park, and its memory is reflected in the design. The west side is illuminated by a ribbon of colourful aluminium "flowers" designed in deliberate contrast to the solidity of the concrete motorway structure. Fluttering 8m up in the air, they draw the visitor through the space evoking the memory of the previous park. The east side consists of a series of planted terraces formed by stone reclaimed from the site demolition works and clad in corten steel. Areas of the uncovered bed rock are incorporated and a rainwater collection system feeds water to the plants covered by the overhead carriageways.

CLIENT
Glasgow Canal Regeneration Partnership (Glasgow City Council, ISIS Waterside Regeneration & British Waterways Scotland)

CONSTRUCTION MANAGEMENT	LOCATION	AREA	PHOTOGRAPHER
Land Engineering	Glasgow, UK	3,850 m²	Dave Morris, rankinfraser landscape architecture

Garscube link 是 Speirs Lock 区重建工程的第一个阶段，这项工程由格拉斯哥运河再生伙伴关系（GCRP），提议。伙伴关系建立在格拉斯哥市政府和 ISIS 沃特赛德重建部门的合作上，并得到了苏格兰 British Waterways 公司的支持。这个项目简单来说就是要呼吁要复兴一个格拉斯哥运河网络和城市中心的连接关系，该连接于 1960 年因 M8 高速公路的建设而被切断。

"凤凰花"设计起着对 Garscube M8 高速公路下的公共区域改造的作用。这处新连接成为 Speirs Lock 周边区域与克莱德运河的出入口，也是格拉斯哥北部大片区域与市中心连接的步行通道。

现有的地下通道环境恶劣：黑暗、嘈杂、肮脏和令人生畏。设计适当地拓宽了面积，将通道变为一个覆盖红色树脂的平滑道路，不会限制它被当作一个倾斜的单独通行的功能。空间的两边根据道路的不同方位进行不同的规划，但在外观上形成了统一。公路修建之前，这里曾经是一个公园——凤凰公园，对它的记忆表现在设计上。西边被一片摆放五彩缤纷铝制"鲜花"的长带区装饰，特地与冰冷的混凝土公路结构形成对比。"凤凰花"在距地面 8 米的空中颤动，让人们通过设计引发对之前公园的记忆。东面包括一系列的绿色梯田，它由从原建筑拆除工程中留下的石料组成，再裹上克尔顿钢固定。裸露的基岩地区被建成雨水收集系统，它浇灌了在高架桥下面的植物。

DESIGNER
Ian McChesney

PHOTOGRAPHER
Ian McChesney

Arrival and Departure

到来与离开

Plymouth is the place where Sir Francis Drake played bowls before defeating the Spanish Armada; the point of emigration for the Mayflower, its Pilgrims and countless others setting sail for America; where Darwin set sail on the Beagle and birthplace of Scott the polar explorer.

Ian McChesney has recently completed a new artwork for the University at Plymouth entitled Arrival and Departure. The commission forms part of the University's 150th year celebrations and completes James Square, a new mixed used development comprising a residential building, refectory and the new Rolle Building by David Morley Architects. The piece comprises two opposing forms redolent of nautical bollards – monuments to journeys. While referring to the many historic voyages starting and ending in Plymouth

they also mark the journeys taken by students, both physical and through time and endeavour. Primarily artworks the forms can also be sat upon, laid upon or danced upon.

To mark the unveiling, students from the University and local schools performed "Moving On", a contemporary dance piece performed around and upon the piece choreographed by Gemma Kempthorne inspired by Plymouth's maritime history and journeys.

Each form weighs two tonnes and measures 1.7m long x 0.9m wide x 0.6m high. They were hand carved from solid blocks of granite from the De Lank Quarry in nearby Bodmin Moor, just over thirty miles away from Plymouth.

普利茅斯是弗朗西斯·德雷克爵士在击败西班牙无敌舰队之前打滚木球的地方、也是清教徒和其他无数的教徒乘着五月花号向美洲大陆航行移民的地方，还是出生于斯科特的极地探险家——达尔文乘着比格尔号起航的地方。

伊恩·麦克切斯尼最近为普利茅斯大学完成了一个新的艺术作品，题为"到来与离开"。这个委任成为该大学 150 周年庆典的一部分，并完成了詹姆斯广场——一个包括住宅建筑、学校食堂和莫利建筑事务所建造的新罗尔建筑的新混合发展项目。

这件作品由两个形式相反、令人怀念的航海护柱组成——古迹之旅，在谈及许多在普利茅斯开始和结束的历史悠久的航程，它们也标记着学生的旅行，包括花体力、花时

间和尽力去做的。主要作品的形式也可以被讨论、被强调或者被否定。

为了纪念揭幕式，来自大学和当地学校的学生进行"转移"，GemmaKempthorne 精心设计的一个当代舞蹈作品在周边进行表演，该作品灵感来自普利茅斯的海洋历史和旅行。

每个雕塑重达两吨，长 1.7 米、宽 0.9 米、高 0.6 米。它们是来自博德名摩尔附近兰克采石场的固体花岗岩的手工雕刻，那儿离普利茅斯仅 48 公里的路程。

ARTIST	CLIENT	LOCATION	PHOTOGRAPHER
Simon Watkinson	Blyth Borough Council	Blyth Market Place, Northumberland, UK	Sean Conboy

Hyperscope

壕沟用潜望镜

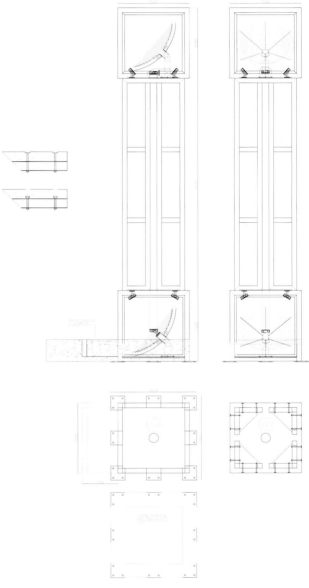

Hyperscope is part of a series of artworks situated in North East England that has seen Simon Watkinson identify how the heritage of certain areas is literally "beneath our feet". The installation highlights the coastal town's rich industrial heritage and forthright regeneration ambitions drawing strongly from the legacy of its shipbuilding and coal mining past including when the Royal Navy used the Port of Blyth to dock their submarines during World War II. Hyperscope is an integrated architectural and lighting artwork, commissioned by Inspire Northumberland that provides a powerful focal point to the design of this beautiful and spacious new square.

壕沟用潜望镜（Hyperscope）是英国东北部一系列艺术品中的一个，它见证了西蒙·沃特金森是如何定义我们脚下这片遗址的。该装置深受造船业和煤矿开采历史的影响，包括二战时期皇家海军在布莱斯港口停泊的潜艇，突显了沿海城镇丰富的工业遗产和坚定的重建雄心。壕沟用潜望镜是综合了建筑和照明的艺术品，由诺森伯兰郡委托，成为美丽宽敞的新广场设计中一个很大的焦点。

Duecentosessantamq

木板艺术

The site of the pavillion was an abandoned tennis court in a private park. As competition requirement, competitors were asked to design a project which could be temporary, easy and quick to build in terms of techniques. There was no restriction about the way to approach to this area by designers and artist.

Since the first sketches, Simone Bossi tries to underline and emphasise what there is in the direct surroundings. He analyses every natural feature close to the project site with the idea to translate them in something artificial: trees became simple wooden board, rocks and gravel became rectangular pieces of stone and a dark lake close by became a square layer of water.

All the natural effects that occour during a day long and seasons suggested the relationship between the choosen elements: filtered sun, changing shadows, water reflections, etc. All the used materials came from local companies. Wooden board was made by pine which was also used for the whole structure.

The result was a layout that let to visitors a freedom to move and discover the area. Even if the pavilion maintains a strong geometry from the top, perspectives, distance and different points of view play an important role by changing the thickness of the intervention and by revealing a new part of the context mirrored in the dark water.

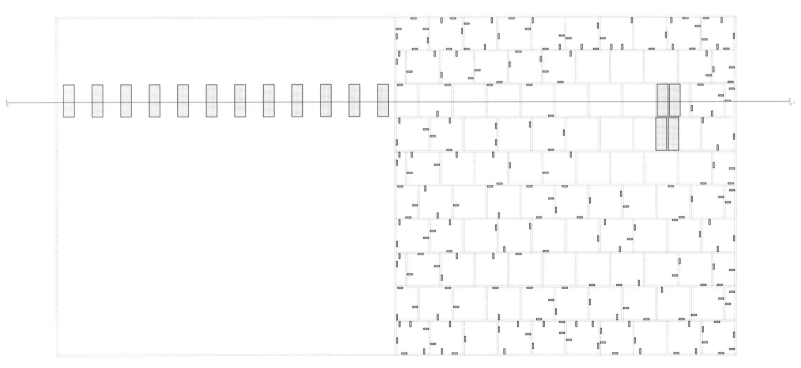

该馆的场地是一个在私人花园里废弃的网球场。正如比赛的要求，选手们需要设计一个技术上临时的、简易而快速搭建的项目。设计师和艺术家们没有被限制到达这一区域的方式。

从第一份草图来看，Simone Bossi 就试图强调并看重在直接的环境中建筑的东西。他分析了接近项目选址的每个自然要素，一边思考着用人工物品来阐释它们。树木变成简单的木板、岩石和砂砾变成矩形块状石头，附近深暗的湖变成一个方形水层。

一天和一个季节所出现的所有自然效果，暗示了所选元素之间的关系：通透的阳光、变化的影子、水面的倒映、等等。使用的材料全部由当地公司提供。松树做成的木板也被使用到整个建造中。

目的是让游客自由走动并发现这片区域的布局。虽然这片区域的顶部保持了一个很明显的几何形状，但是通过改变穿插在其中的木板的厚度和展示深暗水面反射出的部分新镜像，远眺、距离和不同方向的视野起了非常重要的作用。

ARTIST
Simon Watkinson

LOCATION
Sudell Cross, Blackburn, UK

CLIENT
Blackburn & Darwen Borough Council

PHOTOGRAPHY
Sean Conboy

Braid

交织

Visually this stainless steel structure appears to emanate from the floorscape without the need for a mediating plinth. The LED lights in the granite suggest further strands that have been cut, as if the whole structure is in fact a fat fibre optic cable harness emanating from the hard landscaping. The top plates are like the cards used in Jacquard looms for "programming" the design, linking back to the textile industry that characterised the area.

视觉上这个不锈钢结构似乎出自地面而无需中间基底。大理石中的 LED 灯暗示了进一步被切断的股线，仿佛整个结构其实是一个发源于园林建筑工程的粗大的光纤缆线。顶板像为了"编程"设计的提花织机穿孔卡，联系回纺织业、增加了地域特色。

FABRICATION
M-tec.uk

Person Parking

个人停车位景观设计

Person Parking is Springtime's winning entry in the duepercinque design competition in 2009. Duepercinque invites designs that transform public spaces usually dedicated to cars. The only limit is that they will have to be designed to fit into a 2x5m ("due per cinque") parking spot.

The contest has received over 800 projects from all over the world. Ten projects – amongst which the Springtime concept "Person Parking" – were selected for the official duepercinque parade of the Public Design Festival, which was held in Milan from 18th to 26th April 2009. The winners have interpreted public space as a place where people meet, socialise, interact and have found an alternative way to use the parking area.

Person Parking creates an agreeable "parking spot" for people that serves both as a meeting point and a place to rest. Person Parking mimics the visual world of road and road signs and creates a new meaning to the term parking. It also turned out to be a landmark for people to get their photos taken.

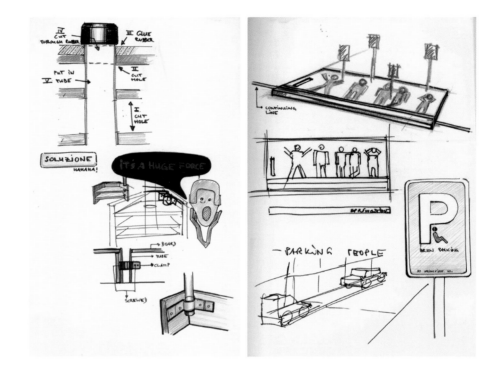

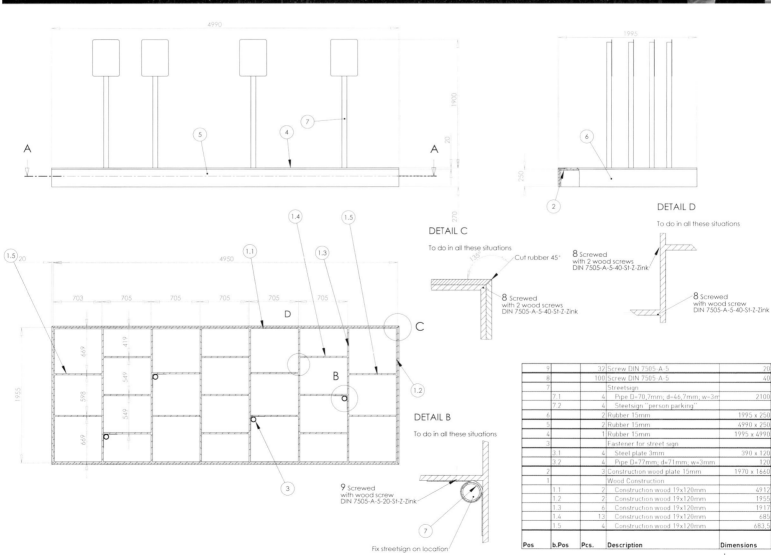

DETAIL C

To do in all these situations

Cut rubber 45°

8 Screwed
with 2 wood screws
DIN 7505-A-5-40-St-Z-Zink

8 Screwed
with 2 wood screws
DIN 7505-A-5-40-St-Z-Zink

DETAIL D

To do in all these situations

8 Screwed
with wood screw
DIN 7505-A-5-40-St-Z-Zink

DETAIL B

To do in all these situations

9 Screwed
with wood screw
DIN 7505-A-5-20-St-Z-Zink

Fix streetsign on location

Pos	b.Pos	Pcs.	Description	Dimensions
9		32	Screw DIN 7505-A-5	20
8		100	Screw DIN 7505-A-5	40
7			Streetsign	
	7.1	4	Pipe D=70,7mm; d=46,7mm; w=3m	2100
	7.2	4	Steetsign "person parking"	
6		2	Rubber 15mm	1995 x 250
5		2	Rubber 15mm	4990 x 250
4		1	Rubber 15mm	1995 x 4990
3			Fastener for street sign	
	3.1	4	Steel plate 3mm	390 x 120
	3.2	4	Pipe D=77mm; d=71mm; w=3mm	120
2		3	Construction wood plate 15mm	1970 x 1660
1			Wood Construction	
	1.1	2	Construction wood 19x120mm	4912
	1.2	2	Construction wood 19x120mm	1955
	1.3	6	Construction wood 19x120mm	1917
	1.4	13	Construction wood 19x120mm	685
	1.5	4	Construction wood 19x120mm	683,5

个人停车位设计是 Springtime 设计公司在 2009 年的 duepercinque 设计大赛中的获奖作品。duepercinque 设计大赛请设计师们对车辆专用的公共区域进行改造。唯一的要求是设计出来的作品要能摆放在 2 米 x5 米的停车位上。

大赛收到来自全世界的 800 多个作品。大赛选取了 10 个参赛作品，其中包括 Springtime 设计公司的"个人停车位"，来参加大众设计节中的 duepercinque

官方游行。游行时间为 2009 年 4 月 18—26 日，地点在米兰。获奖作品诠释了公共区域作为人们会面、社交、互动的场所的内涵，并且发现了停车位的另一种用途。

"个人停车位"创造了一处令人愉悦的"停靠点"，可以聚会交谈，也可以坐下休息。它仿照道路与路标的视觉图像，为名词停车场增添了新的含义。它同时被证明成为一处可以供人拍照的地标建筑。

LITTER BIN

垃圾桶

Bryant Park Litter Receptacles and Recycling System
布莱恩公园的花托垃圾箱循环系统

Nanuk
北极熊牌防护箱

Eclipe
Eclipe 垃圾箱

Phi Litter Bin
Phi 垃圾箱

DESIGNER
Ignacio Ciocchini, IDSA. Industrial Designer

CLIENT
Bryant Park Corporation ,34th Street Partnership

MANUFACTURER
Landscapeforms / Studio 431

Bryant Park Litter Receptacles and Recycling System

布莱恩公园的花托垃圾箱循环系统

The design of the Bryant Park litter receptacles and recycling system was inspired by plants, flowers and nature in general. The product becomes an accent in the park environment rather than a piece of furniture painted with a dark colour and hidden in a corner. The shape symbolises the commitment of the client to dispose trash and materials in a way that minimises the impact on the environment. The nature-inspired design acts as a visual reminder to people that diligent garbage disposal and recycling have a positive impact on the planet.

The design detailing elements and colours were chosen to complement the park's horticultural elements and bluestone pavers. Plain detailing on the petal castings with a fern green colour coating is used for the trash units. Linear detailing relates to the shape of the opening for newspapers and magazines, with a lime green colour coating is used for paper recycling units. The organic detailing relates to the shape of the opening for bottles and cans, with a blue colour coating used for the plastic and metals recycling units.

Material: Cast aluminium petals, cast ductil-iron base, rolled carbon-steel plate structure, spun stainless steel flip-up lid, fabricated stainless steel hinge, stainless steel hardware, and rotationally molded polyethylene liner

DIMENSION
Height: 102cm, Max diameter: 61cm

LOCATION
New York City, USA

PHOTOGRAPHER
Ignacio Ciocchini

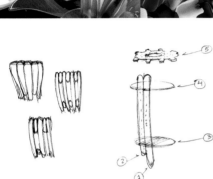

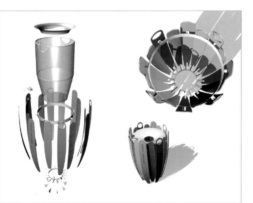

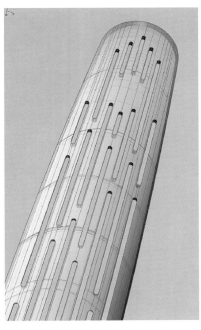
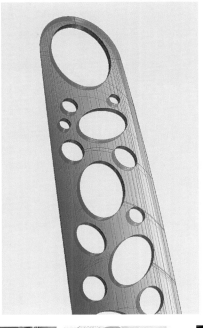

Bottles and Cans

Trash

Newspapers and Magazines

布莱恩公园的花托垃圾箱循环系统的设计灵感来自植物、花朵和自然万物。该产品成为公园的符号，而不是一个藏在角落里的涂上深色的物品。这个造型代表着游客用某种方式处理垃圾和废品的义务，使其最大限度地减少对环境的影响。

受大自然启发的这个设计，对人们勤于处理和回收垃圾产生了视觉性暗示，对我们的地球也有积极的影响。

该设计的细节元素和颜色，选取了可以弥补公园园艺的元素和青石铺材。用于垃圾箱部件的是朴素的花瓣铸片和蕨类植物的绿色涂层。用于废纸回收的部件是线性的设计，它参考了报纸和杂志的开口形状，带着柠檬绿的涂层。不仅如此，有机物细节上参考了瓶罐开口的形状，带着蓝色涂层，它用于塑料盒金属的回收部件。

材料：铸铝花瓣、铸铁基底、碳钢滚板结构、旋转式不锈钢翻转盖、装配式不锈钢铰链、不锈钢器具和旋转型聚乙烯衬垫。

DESIGNER
David Karásek, Radek Hegmon

DESIGN COMPANY
mmcité

Nanuk

北极熊牌防护箱

Anatomically the most temperate design of all, structurally simple but also made of assorted materials – a fine example of smart design, delivering one of the widest ranges of fresh and attractive litter bins. A cylinder and block are the two main parts of the bin's body standing on a central leg. Furthermore, there is the possibility of several types of coatings suitable for general use. While the cylindrical frame is only available coated with pleasing wood lamellas, the block besides them can also be reinforced with bent steel sheet, transparent stretched metal, uncompromising gratings or high pressure laminate (HPL) with smooth interiors. Blocky versions may also be fitted with an elegant cover.

　　防护箱设计简单、内部结构明了，采用多种材料制成。它是设计巧妙的杰作，在众多崭新、美观的箱子里独树一帜。支架支撑的主箱由两部分组成：箱体和嵌板。此外、公司推出多种款式供人们选择。箱体是唯一覆盖光滑木壳的部分，嵌板被弯曲的钢圈、透明材质的隔板、坚硬的钉子以及高压胶合板（HPL）牢牢地固定在箱子内部。箱型结构被装饰上了华丽的外壳。

DESIGNER
David Karásek, Radek Hegmon

DESIGN COMPANY
mmcité

Eclipe

Eclipe 垃圾箱

A sequence of elegant ellipses creates an outer shell that looks like finely modelled ribbing, which determines the characteristic impression of this distinctive and at the same time natural-looking litter bin. The litter bin is available in a light-weight version based on a central leg and a less conventional version with a steel panel base; both models are available with or without a cover. Now, there is also an elegant and practical version designed to be affixed to a wall.

优雅的连续椭圆形创造了一种看起来像精美的仿制螺纹的外壳，它决定了这种独特的而看起来又浑然天成的垃圾箱的特征印象。垃圾箱的一种样式是重量轻的，以中央支柱为基底，另一种样式是以钢板为基底的，显得不那么的传统。两种样式都可以搭配，或者不搭配盖子。现在，还有一个优雅又实用的样式，是安置到墙上的。

DESIGNER	PRODUCTION AND COMMERCIALIZATION	MATERIALS	AVAILABLE COLOURS
Diana Cabeza	Estudio Cabeza	Round section iron rod, welded and painted	Orange, light blue and green

Phi Litter Bin

Phi 垃圾箱

It is a simple and efficient litterbin, with low maintenance requirements.

该垃圾箱结构简单，性能高效、无需过多地维护保养。

INSTALLATION

Free standing, fixed to ground or staked on grass

SIZE

bigger Ø 0,74 m, smaller Ø 0,37 m, h= 0,87 m

LIGHTING
照明

LANDSCAPE ARCHITECT	ARCHITECT	CLIENT	LOCATION	AREA
b:d landscape architects ltd	Hawkins\Brown	London South Bank University	London, UK	2,345 m²

LSBU Student Centre

伦敦南岸大学学生中心

Completed in December 2012 the new Student Centre at London South Bank University represents the first "anchor" within London South Bank University's Estate Strategy to revitalise the campus with enhanced public realm improving connections and legibility between campus and urban fabric of the area.

The landscape design creates a more attractive public realm with an "outward-looking" campus with a very special sense of place with life and activity to the enhanced Kell Street and Public Square on Borough Road. Using a high-quality palette of materials inspired by the history and heritage of the site with semi-mature tree planting and rich sensory shrub and herbaceous perennial beds to promote and enhance biodiversity for urban ecology.

Drawing inspiration from history, most notably David Bomberg – one of the Whitechapel Boys – who taught in the 1960s in the surrounding buildings, the aim for the rejuvenated public realm gateway is to be a dynamic "carpet" of quality granite setts that draws out the strong geometry of the undercroft structure into the landscape.

This rigidity of structure is fractured and exploded with highlights of linear inground lighting, contrasting granite planks and granite and timber benches that draw pedestrians in from Borough Road but also allow for colonisation and use throughout the day.

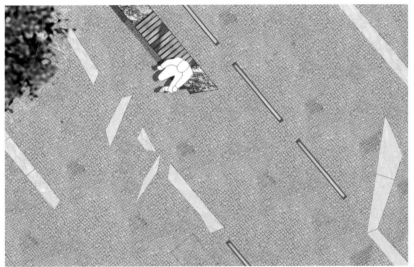
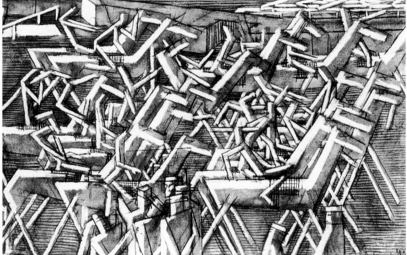

　　伦敦南岸大学学生中心于 2012 年 12 月竣工，它是伦敦南岸大学地产策略系列中的第一"站"，该策略旨在扩大公共区域，使校园重获生机，同时改善校园与该区域城市构造之间的联系，提高辨识度。

　　这一景观设计建造了一个更具吸引力的公共区域，其中包括一个外向型的校园，使伯勒路上扩建后的凯尔街道和公共广场蒙上了一层独特的色彩，富有生机和活力。受历史与文化的启发，该区域选用了高品质格调的材料，同时为了提升城市生态环境的生物多样性，种植了半成熟植物、大量灌木林以及多年生草本植物。

　　很显然，大卫·邦勃格从历史中获得灵感。他在怀特查佩尔区长大，于 19 世纪 60 年代在周围地区教书。该设计利用一块块高品质的小方块花岗岩建造出一条颇具活力的"地毯"式通道，地下室构造中坚固的几何图案延伸出来呈现在"地毯"上，与风景融合在一起，从而实现复兴公共领域通道的目标。

　　这种构造并不刻板，线型埋地灯尤为惹眼，与花岗岩木板、花岗岩和木质长椅形成鲜明对比，不仅将行人从柏勒路吸引至此，还考虑到了集群现象，而且全天都能使用。

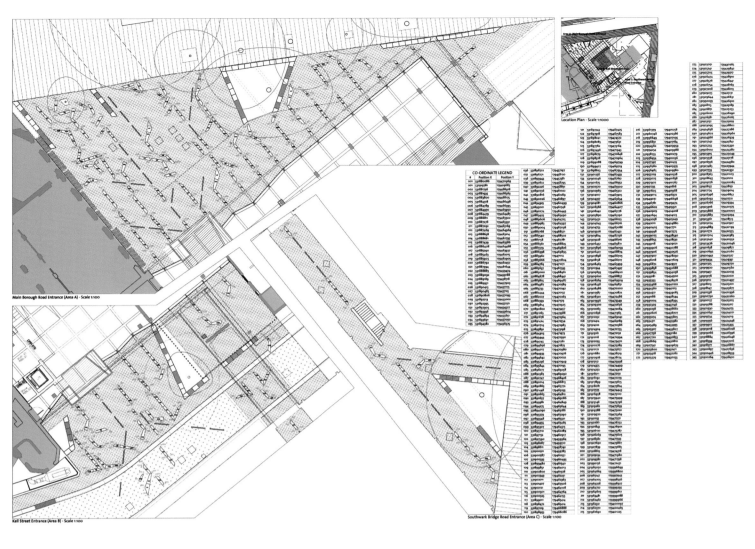

Main Borough Road Entrance (Area A) - Scale 1:100

Kell Street Entrance (Area B) - Scale 1:100

Southwark Bridge Road Entrance (Area C) - Scale 1:100

Location Plan - Scale 1:1000

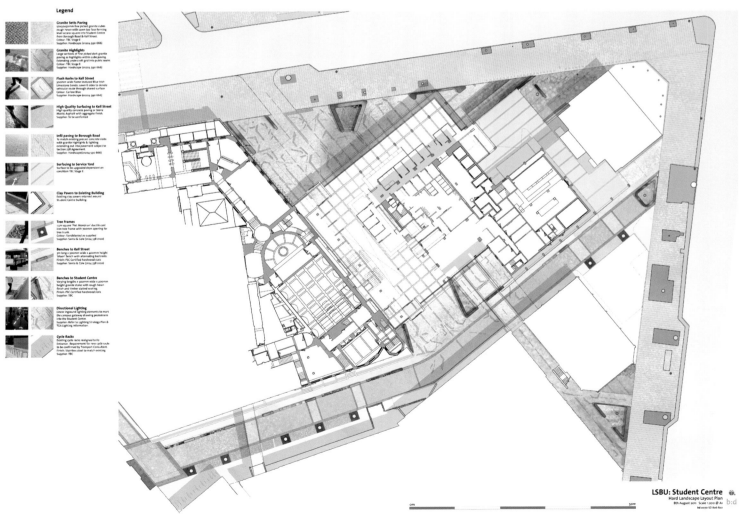

Legend

LSBU: Student Centre
Hard Landscape Layout Plan

b:d

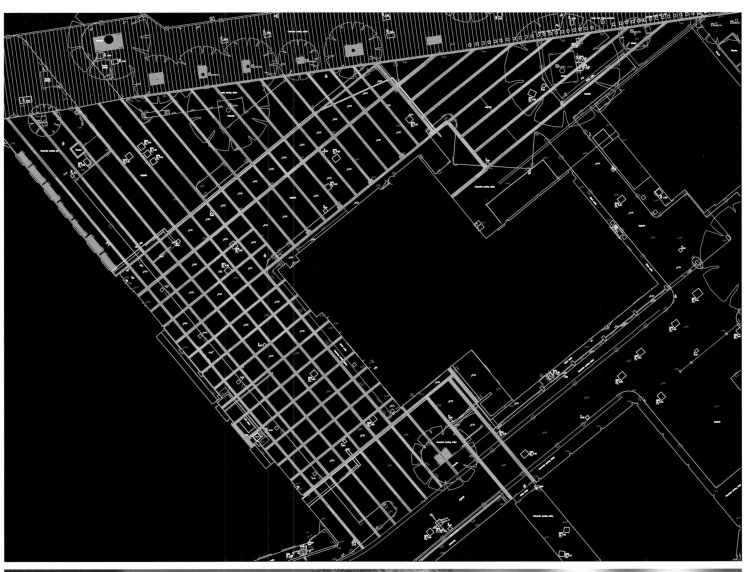

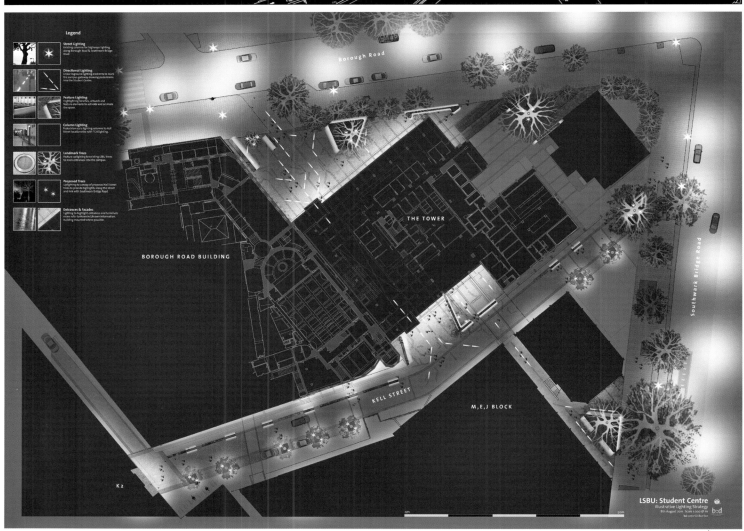

Legend

Street Lighting

Directional Lighting

Feature Lighting

Column Lighting

Landmark Trees

Proposed Trees

Entrances & Facades

Borough Road

THE TOWER

BOROUGH ROAD BUILDING

KELL STREET

M,E,J BLOCK

Southwark Bridge Road

K 2

LSBU: Student Centre
Illustrative Lighting Strategy

b:d

DESIGN COMPANY
Pensa

PHOTOGRAPHER
Pensa

StreetCharge

街边充电设施

StreetCharge is a solar-powered installation that provides charging, lighting and a place to rest. It is a simple and elegant solution to a common problem. In the urban environments, people are often in need of a quick place to recharge the mobile devices. Similarly, they often need a brief respite to recharge themselves as well.

The concept of a solar-powered street-side charging station built onto street signs grew out of the designers' interest in creating a durable and self-sustaining solution which could not only live off the grid, but be built from reclaimed and local materials that already existed in the urban environment. Too often they see elements added to their built environment with no regard to their surroundings. They wanted to create an element that could be seen as functional, quiet, and complementary, and had the potential to be implemented on a larger scale.

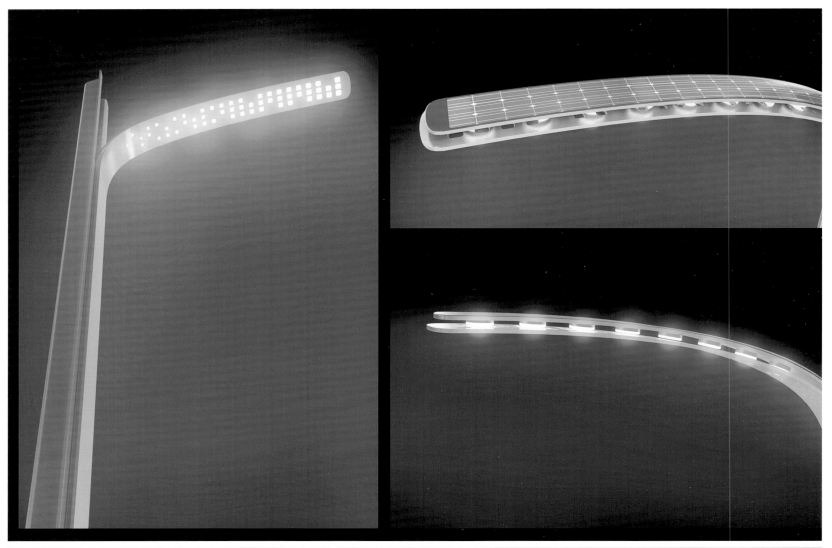

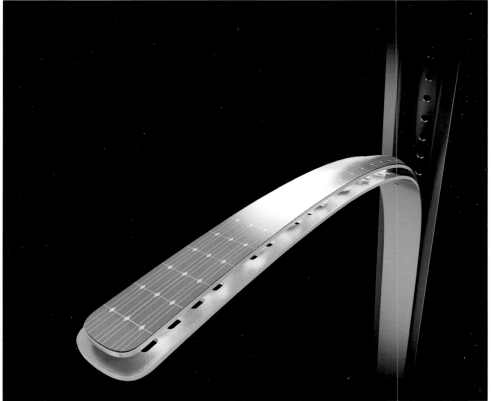

　　街头充电设施是一种太阳能驱动的设施，提供充电、照明和休息的地方。它是对一个常见问题的一个简单而优雅的解决方案。在城市环境中，人们经常需要一个给移动设备快速充电的地方。同样，他们往往需要短暂的喘息机会也给自己充充电。

　　在街道指示牌上安装一个街边太阳能充电站的概念来自于设计师希望创造一个持久的解决方案，不仅依靠电网，更依靠回收的已经存在于城市环境中的材料。

　　他们经常看到没有考虑到人们周围的环境，很多元素就被添加到环境建设中去了。他们想创造一种元素，可以看作是功能性的、安静的、补充性的、并有更大范围内实施的潜力。

DESIGNER
David Karásek, Radek Hegmon

DESIGN COMPANY
mmcité

Elias

Elias 柱体

This slim and geometrically shaped bollard has a longitudinal cavity whose appearance changes depending on the angle from which it is viewed, ranging from a dark shade to full transparency. This simple block with extreme cannelure is very contemporary and yet almost classic in character suitable for any environment. Upon request, an embossed municipal emblem made of zinc alloy can be embedded into the body of the bollard (available upon request). What is dark during the day is lit up at night. Low-voltage LED technology can also be installed inside the bollard and serves as a decorative lamp with the same design as the rest of the Elias bollards.

　　这根细长的几何形柱体是纵向中空的，其外观的变化取决于从哪个角度去看它。变化范围从黑影直至完全透明。简明的柱身搭配极度的纵槽、这样现代而又经典的特征，适应任何环境。根据要求，由锌合金制成的压花市徽可以嵌入柱体（可根据要求做）。　白天未点亮的黑暗处在晚上被点亮。也可以把低压的 LED 科技品安装进柱体里，并作为 Elias 柱体其他相同设计的一种装饰灯。

Rotermann Quarter

罗特曼大厦的配套公共设施

The recurrent theme of Rotermann Quarter is creation of atmosphere characteristic to trade yards. The leading element is a stylised package as a seat, luminaire and streetlights.

Distinctively prominent in night and day, the rotomoulded plastic seats Quantum enrich the modern city environment, appearing light, timeless and temporary at the same time. These benches with integrated LED lights give the nightly city space an unforgettable shine and a new face. Thanks to its resistant plastic material, it defies insidious weather conditions and graffiti artists – pressure washing makes everything clean again, turning Quantum benches back into a nice place where to rest one's feet. It is also suitable for garden and patio use –the bench and lighting in one.

　　罗特曼大厦配套设施这一反复出现的主题是为交易码头独特的氛围而创作的。其主导元素是座位、照明设施和街灯的配套。不论在白天或黑夜，塑料座椅总是鲜明而独特，为城市增添了现代气息，透着光亮，同时显得永恒又短暂。

　　这些长椅都内置了 LED 灯，到了晚上，它们让城市的夜景换上一张新面孔，令人难以忘怀。得益于其耐热性塑料材质，它与恶劣的天气状况和街头涂鸦抗争，只要高压清洗，就能使其焕然一新，然后让椅子回归原位，再次造福市民。

　　这些设施同样适用于花园、庭院的使用——包括椅子和照明灯柱。

DESIGNER
Margus Triibmann

DESIGN COMPANY
Keha3

MATERIAL
aluminum Glass - polycarbonate, LED

Branch

节支路灯

Branch is an example of how Keha3 LED profiles can be used in street luminaires. Branch is a dynamic luminaire concept of light traffic roads and yard areas the different shapes of which are inspired from the tree crowns and flower heads. With combinations of the number of branches of the luminaire and power of LED elements located in profile and the secondary optics they can create an endless number of versions in the field of street luminaires in lighting technological as well as visual-aesthetic category. The number of elements in the crown can be varied according to the design or goals for lighting solution. Single elements of the luminaire can be freely directed with the joints, and hence they can make light pattern on the street as well as achieve required evenness of the light, and at the same time illuminating the surrounding buildings and objects. Branch luminaires are suitable for special design as well as standard design streetlight poles.

　　节支路灯是 KEHA3 公司的 LED 路灯如何用于街道照明的一个例子。节支路灯是点亮交通道路的一个动态灯具，其设计理念来自于形状各异的树冠和花朵。节支路灯通过外形和二次光学元件在发光体和 LED 元件功率的不同组合，可以将街道照明改变成无数种照明方式，提供了技术照明以及视觉审美功能。树冠的元件数量可以根据设计方案或者照明方案来改变。发光体的单一元件可以在支点上任意定向，因此，它们可以在街道上创造各种灯光效果，也可以达到人们所需的光亮均匀度，同时又能给周围的建筑和物体打上灯光。节支照明路灯适用于特殊设计和标准设计的街道灯杆。

ARTIST
Simon Watkinson

LOCATION
Telephone House, High Street Southampton, UK

CLIENT
Linden Homes

PHOTOGRAPHY
Rod Varley

Telephone House High Street Southampton

南安普顿高街电话亭

With a brief to create a new public space outside the converted Telephone House, Simon created a space that took the grid as his starting point. Archeological references and sugar making provided some of the key imagery in this scheme with a series of bespoke granite lighting cones filled with toughened glass providing a lead to the building's entrance.

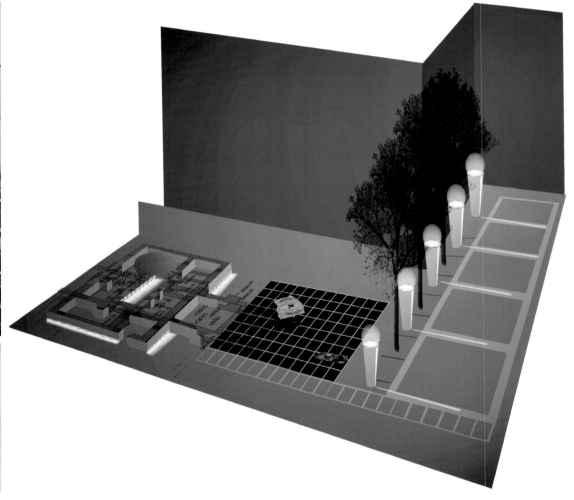

简单来说是要在可改变的电话亭外创建一个新的公共空间，设计师西蒙以电话亭为出发点来设计这片空间。考古资料以及制糖工艺为该方案提供了一些重要意向，钢化玻璃定制的花岗岩照明锥立在建筑物的入口处提供指引。

DESIGNER
Javier Machimbarrena

SKA Streetlamp

SKA 街灯

In the relationship between the human being and the environment, the designers absorb the sensations from the exterior like in an osmosis process, being their interior undoubtedly affected by the vision of the environment. It is their responsibility as designers and architects to create items which are faithful with the environment, not forgetting that the landscape aesthetics and its elements constitute a public and trans-generational property.

The SKA streetlamp, a project for the Urban Furniture Company ONN Outside promotes the importance of form and material in the creation of high-quality products.

Its physiognomy, with the curves and organic shapes depurated to the minimum post-geometric expression, contemporary and differentiated, allows its adaptation to all possible environments causing a minimum impact.

The details in its structure and materials makes them also to perceive its value and consideration when projected and constructed, being plausible that sensibility that the designers consider essential in every action or production made by the human being in every situation.

Materials and specifications: Galvanised steel post with white polyester paint finish; aluminium injection lighting with a polyester paint; reflector made of printed aluminium, with a 4mm tempered glass; light bulb: Cosmopolis Cosmo White by Philips: 40W and 60W; H 400/600cm, W 30cm, D 55cm

在人类与环境的关系上，设计师从外界吸收灵感，就像渗透的过程，它们的设计无疑是受到周围环境启发的。他们作为设计师和建筑师有责任创造忠于环境的事物，但也不能忘记景观的美学价值以及它构成跨时代公共财产的所需元素。

作为城市家具公司 ONN Outside 的一个项目，SKA 街灯提升了创造高质量产品中的形式和材料的重要性。

它的外观：弧度线条和有机的形状，简洁到用最少的几何表达、当代设计并具有鲜明特色，这使得它对环境产生很小的影响，可以适用于任何环境。

一旦建造，它在结构设计和材料选用上的细节又可以让人们感受到它的价值和精心设计。设计师认为要使设计合理、感性是人类在任何情形下的行动和生产的基本考虑。

材料和规格

白色聚酯漆的镀锌钢柱。镀铝和聚酯漆照明。印刷铝反射器，具有 4 毫米钢化玻璃。灯泡：大都市 COSMO 白色飞利浦：40W、60W。H 400 / W 30cm；D 55cm 600cm

COMPANY	MATERIALS	LOCATION	MODEL	PITCH
AHL Linghting Group	Steel wire	Hebei, China	AHL-L4565	200mm

Information Tree

信息树

Achieve soft lighting effects using small LED pixels, with total quantity of 1,980 pcs.　　达到了柔和的灯光效果，使用了总数为 1980 个小型 LED 灯。

BUS STOP
公交站台

DESIGNER
LYVR, Lysbeth de Groot-de Vries

ARCHITECTURAL DRAWINGS
Johan de Groot

LIGHTING
Philips

CLIENT
Municipality of Groningen, Department ROEZ

Busstop Park+Ride Citybus

城市公交停车转乘站

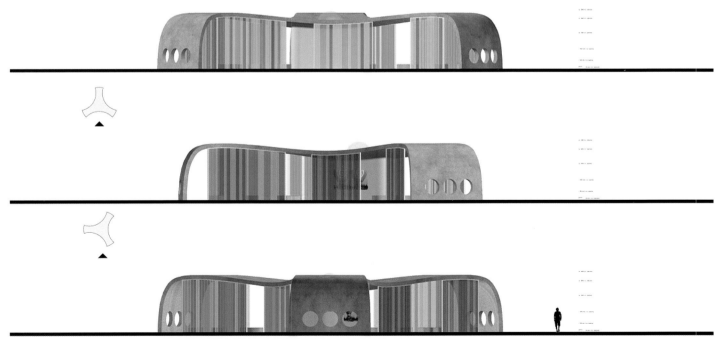

ENGINEERING
Ingenieursbureau Gemeente Groningen (IGG)

LOCATION
Hoogkerk, The Netherlands

PHOTOGRAPHER
Johan de Groot

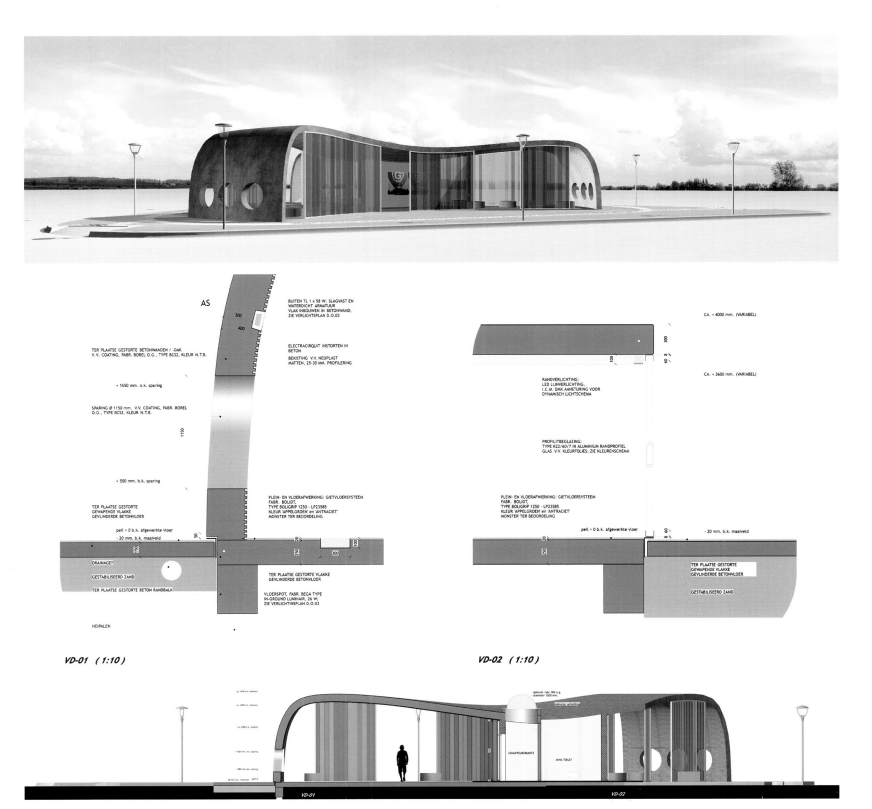

The site is located at the south side of the A7 near Peizermade on the border of the province of Groningen and Drenthe. The P+R offers space for 600 cars, bicycles and offers various options for transfer from county and city buses.

The waiting room is located on the edge of the parking and the bus lane, directly offering access to the Central Station of Groningen. The waiting area is a partially enclosed public space that consists of a square and a building with waiting facilities.

The triangular square is characterised by a pattern of lily leaves that extends to inside the building. The large lily leaves are a reference to the countryside with many ditches, the area where the building is located. The square is slightly elevated. Handicapped access the square by ramps and fysical guidelines are integrated in the surface.

The main structure of the building is a domed concrete shell that is incised on three sides. The form knows no right angles, but is composed of circles, arcs and spheres. Inside is a triangular metal core which includes a disabled toilet, a utility room and a driver room. The central core is also part of the main structure that is self-supporting. The shape of the building is a clear striking sculpture in the landscape. The openness of the building creates a public safety. Lighting also plays an important role. The recessions are partly filled with glass sections, for which is chosen three different colour schemes, which refer to the three directions from where the buses come together. At the same time, these glass walls also give shelter from wind and rainfall. At the evening the building is lit in several ways directly and indirectly so that it looks like a multicoloured lantern light. Inside the room round concrete pouffes offer passengers a dry and sheltered waiting experience. The timetable of the buses are visible within and outside the waiting room via an integrated digital information system.

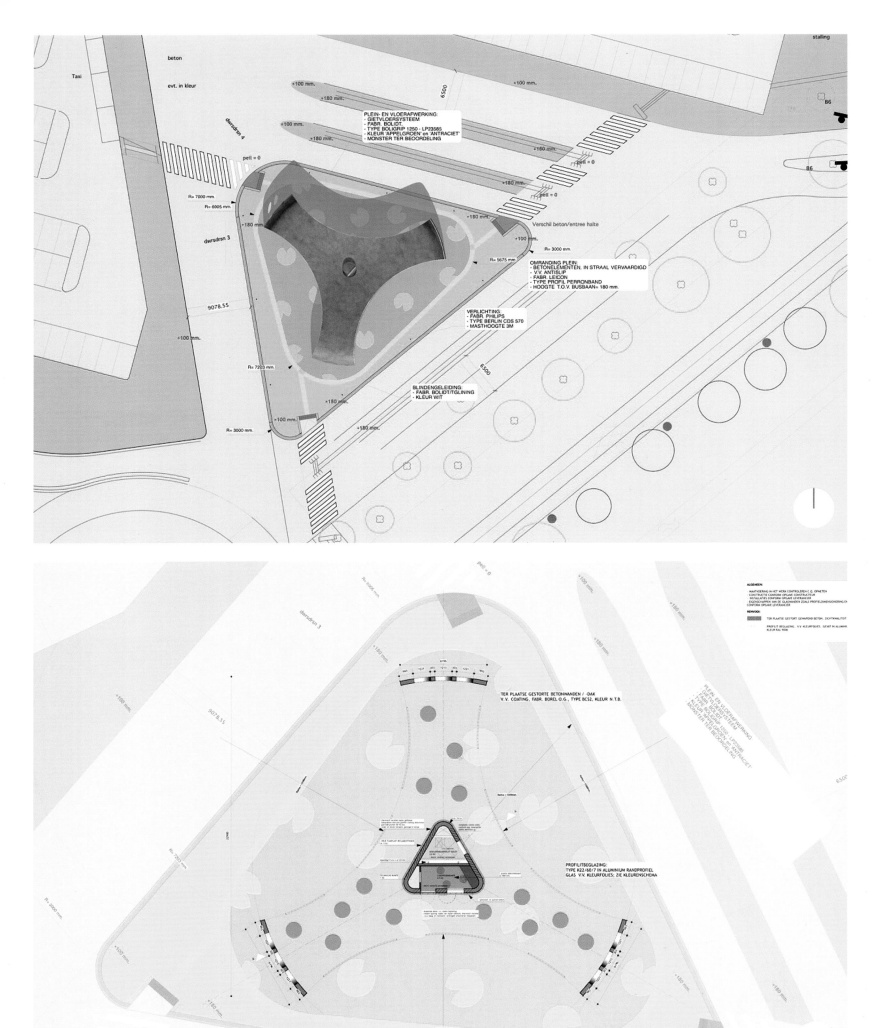

PLEIN- EN VLOERAFWERKING:
- GIETVLOERSYSTEEM
- FABR. BOLIDT.
- TYPE BOLIGRIP 1250 - LP23585
- KLEUR 'APPELGROEN' en 'ANTRACIET'
- MONSTER TER BEOORDELING

OMRANDING PLEIN:
- BETONELEMENTEN, IN STRAAL VERVAARDIGD
- V.V. ANTISLIP
- FABR. LEICON
- TYPE PROFIL PERRONBAND
- HOOGTE T.O.V. BUSBAAN= 180 mm.

VERLICHTING:
- FABR. PHILIPS
- TYPE BERLIN CDS 570
- MASTHOOGTE 3M

BLINDENGELEIDING:
- FABR. BOLIDT/TGLINING
- KLEUR WIT

Verschil beton/entree halte

TER PLAATSE GESTORTE BETONWANDEN / -DAK
V.V. COATING, FABR. BOREL O.G., TYPE BC52, KLEUR N.T.B.

PROFILITBEGLAZING:
TYPE K22/60/7 IN ALUMINIUM RANDPROFIEL
GLAS V.V. KLEURFOLIES; ZIE KLEURENSCHEMA

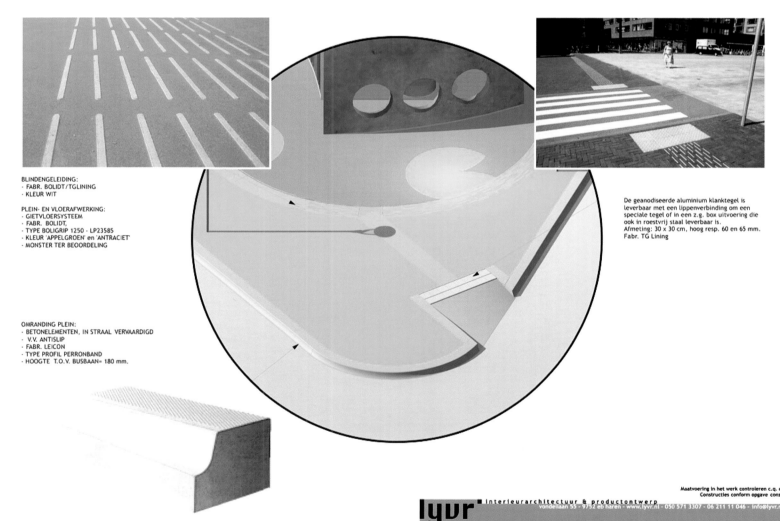

BLINDENGELEIDING:
- FABR. BOLIDT/TGLINING
- KLEUR WIT

PLEIN- EN VLOERAFWERKING:
- GIETVLOERSYSTEEM
- FABR. BOLIDT,
- TYPE BOLIGRIP 1250 - LP23585
- KLEUR 'APPELGROEN' en 'ANTRACIET'
- MONSTER TER BEOORDELING

OMRANDING PLEIN:
- BETONELEMENTEN, IN STRAAL VERVAARDIGD
- V.V. ANTISLIP
- FABR. LEICON
- TYPE PROFIL PERRONBAND
- HOOGTE T.O.V. BUSBAAN= 180 mm.

De geanodiseerde aluminium klanktegel is leverbaar met een lippenverbinding om een speciale tegel of in een z.g. box uitvoering die ook in roestvrij staal leverbaar is.
Afmeting: 30 x 30 cm, hoog resp. 60 en 65 mm.
Fabr. TG Lining

Maatvoering in het werk controleren c.q. opmeten
Constructies conform opgave constructeur

lyvr ■ interieurarchitectuur & productontwerp
vondellaan 55 - 9752 eb haren - www.lyvr.nl - 050 571 3307 - 06 211 11 046 - info@lyvr.nl

tekening: Materialisering Oprit / Toegang plein & Blindengeleiding

该地点坐落在 A7 的南面，位于格罗宁根与德伦特省边界，毗邻 Peizermade。城市公交停车转乘站内可以停放600辆汽车和自行车，此处汇集多条城乡间转乘线路。

等候室位于停车场和公共汽车专用车道的边缘地带，从这里可以直接进入格罗宁根中央车站。等候区是一个半封闭的公共区域，包括一个广场和一栋建筑，内设等候设施。

广场呈三角形，百合花花瓣图案一直延伸到建筑内部。巨大的百合花花瓣的原型来源于建筑所在地的许多乡下沟渠中生长的百合。该广场地势微高。广场上设置了残疾人专属坡道，并在坡道表面铺设标线，以便起到指引作用。

该建筑的主要结构是一个三边雕刻的拱形混凝土壳。这座建筑形状上并没有直角，取而代之的是圆形、弧线和球体。内部是一个三角形的金属芯，内设供残障人士使用的卫生间一间、杂物室一间以及驾驶员休息室一间。项目的主建筑结构独立自持，中央核心位于其中。建筑的形状引人注目，明朗清晰。其开放性保障了公共安全。灯光同样扮演了重要的角色。部分墙壁凹进的地方镶有玻璃立面，并选择三种不同的配色方案，代表车辆驶来的三个方向。与此同时，这些玻璃墙同样起到防风挡雨的作用。在夜里，或直接或间接，建筑以不同的方式被点亮，看起来就像是一个多彩的灯笼。房间内部，人们可以躲避风雨、坐在圆形混凝土大座椅上休息。综合数字信息系统提供公交车时刻表，在候车室内、外均可以看到。

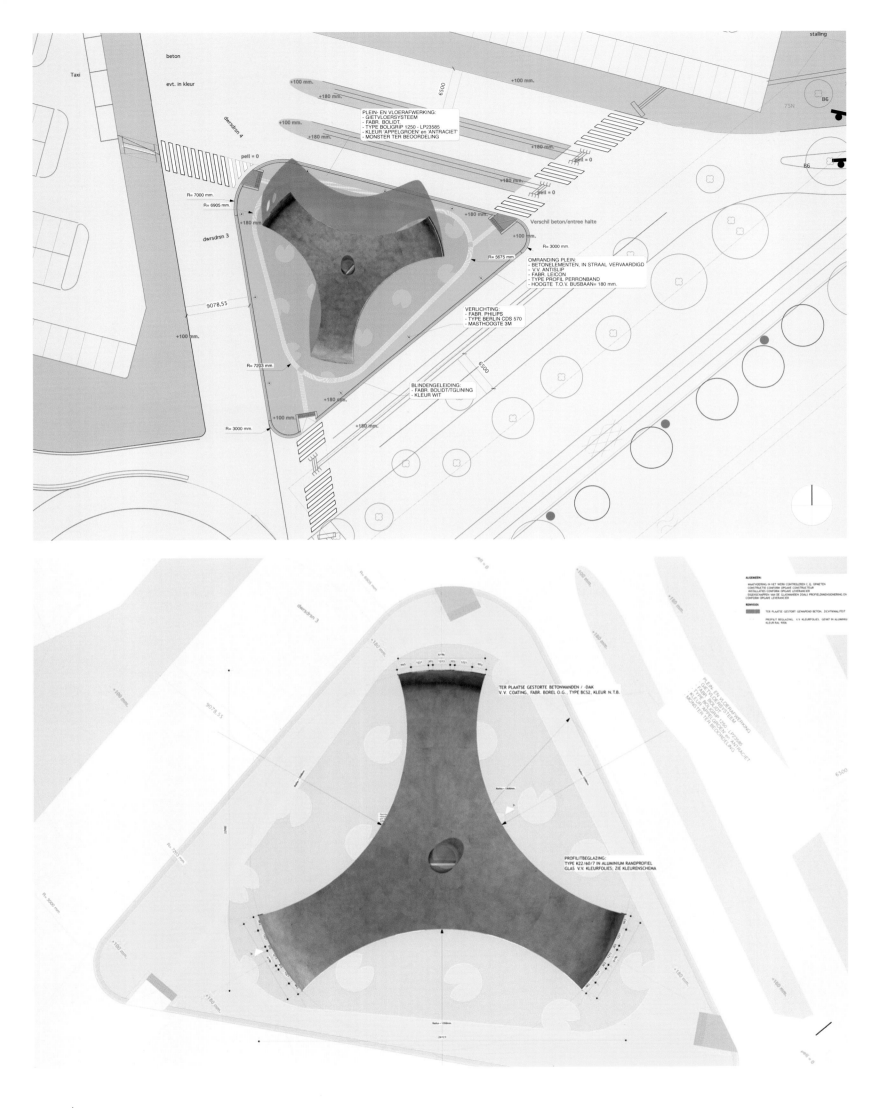

PLEIN- EN VLOERAFWERKING:
- GIETVLOERSYSTEEM
- FABR. BOLIDT.
- TYPE BOLIGRIP 1250 - LP23585
- KLEUR 'APPELGROEN' en 'ANTRACIET'
- MONSTER TER BEOORDELING

Verschil beton/entree halte

OMRANDING PLEIN:
- BETONELEMENTEN, IN STRAAL VERVAARDIGD
- V.V. ANTISLIP
- FABR. LEICON
- TYPE PROFIL PERRONBAND
- HOOGTE T.O.V. BUSBAAN= 180 mm.

VERLICHTING:
- FABR. PHILIPS
- TYPE BERLIN CDS 570
- MASTHOOGTE 3M

BLINDENGELEIDING:
- FABR. BOLIDT/TGLINING
- KLEUR WIT

TER PLAATSE GESTORTE BETONWANDEN / -DAK
V.V. COATING, FABR. BOREL O.G., TYPE BCS2, KLEUR N.T.B.

PROFILITBEGLAZING:
TYPE K22/60/7 IN ALUMINIUM RANDPROFIEL
GLAS V.V. KLEURFOLIES; ZIE KLEURENSCHEMA

DESIGN COMPANY
Atelier Cité Architecture

LOCATION
Allier, France

Couverture de la Gare Routiere de Vichy

维希公共汽车站覆盖区

A station between the spa town and its neighbourhoods activities, Vichy Train Station was opened in 1865. Over time, the station transported its uses. The central pavilion, located in the axis of the trident, is no longer the main entrance. The redevelopment of the station square Vichy proposes to retain an overall geometric design in a simple and rational way for many complex functions of this space.

A monumental court, where the presence of the plant is strong, is arranged in the axis of boulevards access for pedestrians, bicycles and automobiles. It restores the relationship between the station area and the spa town with its planted walks and parks system. A more functional court, hosting a large part of the services of travel and transport interchange (bus station, taxi station locations for rental cars), is fitted with reference to the active and commercial district. The first concern of this project is to organise in a simple geometric design with different functions and uses of the site.

维希火车站于 1865 年正式对公众开放，位于温泉小镇和其邻近活动区之间。随着时光的流逝，车站的用途发生了改变。中心凉亭位于三叉线的中轴上，已经不再是主入口。对于车站广场的再开发，维希建议保持一个整体规划设计，采用简单合理的方式保留此处许多复杂的功能。

庭院内植物生机勃勃，地址选在林荫大道的中轴线上，供行人、自行车以及汽车行驶。利用由植被绿化的道路和园林系统重建车站区域和温泉小镇之间的关系。在活动区和商业区设置了一个功能更强大的庭院，很好地为旅游和交通换乘（公共汽车站、供租赁车辆的出租车站）提供服务。该项目首先考虑的是建立一个简单的规划设计，使该处具有与众不同的功能和用途。

DESIGNER
David Karásek, Radek Hegmon

DESIGN COMPANY
mmcité

Aureo

Aureo 休憩处

一系列的休憩处有两种屋顶材料的选择，并往往选择大的材料。这使你能够选择当地最好的休憩处。聚碳酸酯或玻璃制成的拱形或平形的屋顶，用许多有趣的细节给单一的结构施压。包括在后壁的支撑系统允许使用无侧壁的样式，狭小的空间尤为方便。保证高抗破和耐腐蚀。

A range of shelters with two roofing options and large material selectivity. This enables you to choose the best shelter for the locality. An arched or flat roof made of polycarbonate or glass bears upon the pure structure with many interesting details. A supporting system included in the rear wall allows you to use the version without side walls, especially convenient for narrow spaces. Guaranteed high resistance to vandalism and corrosion.

DESIGNER
Choi Soon-yong

DESIGN COMPANY
The Ground Studio

LOCATION
Sowol road's bus stop of Namsan Library, South Korea

Pictorial Montage

画报蒙太奇

This project is located at the entrance of Namsan Sowol Broadwalk, on the ridge of beatuiful Namsan Mountain in Seoul. This site has characterised the boundary, which is intersected in nature and city landscape.

The design concept is "Pictoral Montage" (art canvas) which is to show the value of a natural landscape throughout architectural transparency.

For this concept, The Ground Studio designed perceptual frame "Pictorial Montage" as spatial experience as Mt. Namsan's natural landscapes and the memory of time for the urban (Seoul).

Art shelter is consisted of three frames, which are wooden bench frame for the static rest, exposed concrete gate frame for the dynamic flow of bus passengers and steel canvas frame of curtain type for the recognition of natural value.

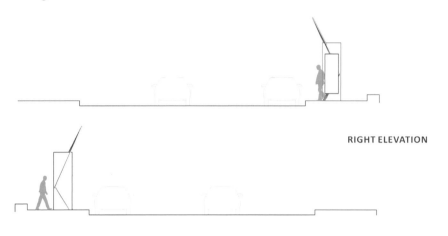

RIGHT ELEVATION

LEFT ELEVATION

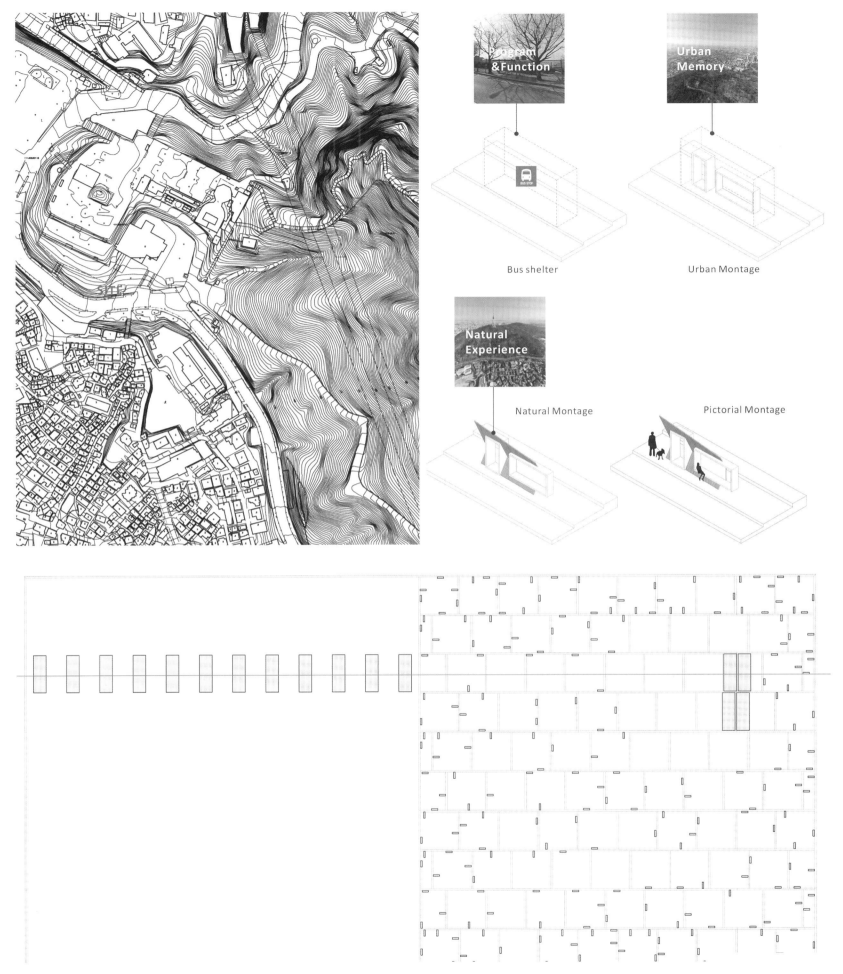

Bus shelter

Urban Montage

Natural Montage

Pictorial Montage

该项目位于首尔南山山脊沿湖栈道的入口处。它将自然风光与城市风景分割开来，形成了独具特色的分界线。

该设计的理念是"画报蒙太奇（艺术油画）"，目的是透过建筑凸显自然景观的价值。

在这一理念的指导下，设计师们设计出了"画报蒙太奇"这一感知框架，展现南山自然风光的空间体验和对城市（首尔市）的回忆。

艺术凉亭由三个构架组成，分别是木质长凳、无遮蔽的混凝土门框和幕式钢制油画内框。在这里，游客可以安静地坐在木质长凳上休息，公交乘客可以进出穿行于混凝土门框，而钢制油画框则突出了自然的价值。

SECTION-1

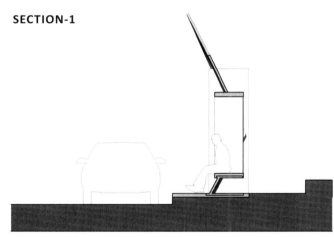

SECTION-2

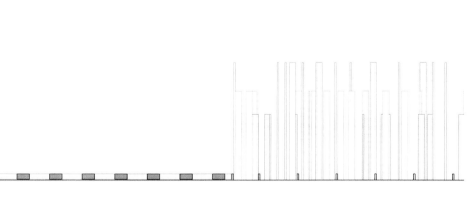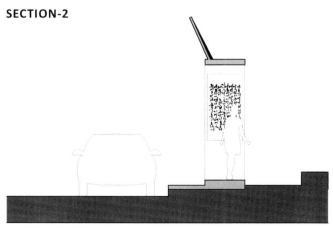

TAXI Station

出租车站

TAXI Station offers two size options as "T1" and "T2" to better adapt to physical conditions and space requirements. "I" profile steel supports provide a graphical identity besides functioning as construction elements. Using power LED lighting, yellow coloured letters turn the station into an urban design symbol at day and night.

The design has an active billboard for advertising purposes which may add to running costs. In addition, the metal shell collects rainwater which is used in restroom and provides space for a green wall. Aiming to answer contemporary city life expectations, TAXI! Station is an easy-to-build and -maintain construction with a modern taste.

T1 size has 12.8-square-metre closed space including waiting lounge, office, kitchenette, storage and WC. T2 size has a total of 32-square-metre closed space including 12-square-metre waiting lounge with a capacity of 10 passengers, office space, changing room with personal closets for drivers, resting space, kitchen, storage and WC. Installation layer is hidden below ceiling and includes multiple air conditioners and water reservoirs. Layer is accessible through three façades with removable panels. Compact treatment is used for toilets.

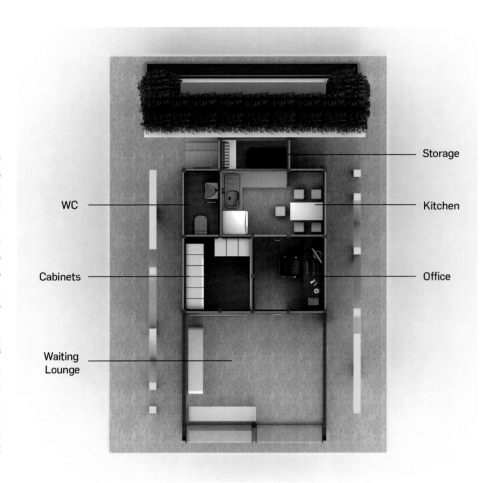

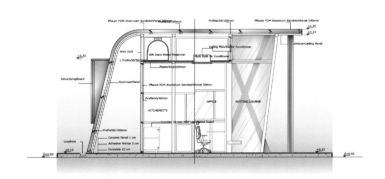

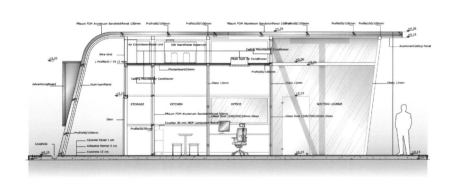

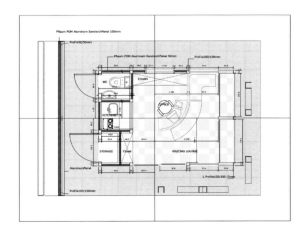

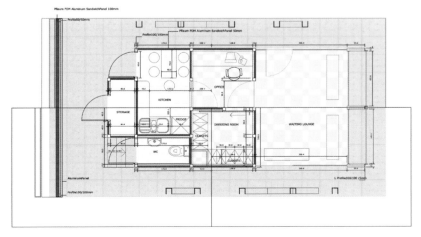

出租车站根据不同的物质条件和空间需求共有 T1 和 T2 两种尺寸可供选择。"I"形钢架作为建筑元素在发挥作用的同时，还是形象生动的标识。无论是白天还是夜晚，通过大功率 LED 照明灯，黄色的字母将车站变成城市的设计标志。

该设计使用了生动的广告板，以此来达到宣传目的，不过同时也增加了运行成本。另外，金属的外壳可以收集雨水，这些雨水可以在洗手间循环使用，同时金属壳也为绿色植生墙提供了空间。出租车站便于搭建，为建筑保留了现代气息，满足了对当代城市生活的期望。

T1 尺寸的出租车站是 12.8 平方米的密闭空间，包括候车室、办公室、小厨房、储物间和卫生间。T2 尺寸的出租车站是 32 平方米的密闭空间，包括能容纳 10 名乘客的 12 平方米候车室、办公空间、配有个人衣柜的司机换衣室、休息区、厨房、储物间和卫生间。设备夹层藏在天花板下面，包括多功能空调和储水池。该夹层使用了三面外墙的可拆卸镶嵌板，易于进入。卫生间的装修则相对简洁。

DESIGNER
Arch. Diana Cabeza, ID Martín Wolfson

GRAPHIC DESIGN
GD Gabriela Falgione

DESIGN TEAM
ID Diego Ross

CLIENT
Government of the City of Buenos Aires

Metrobus Bus Shelters Design

公共巴士候车亭设计

The Metrobus (rapid-transit bus) bus shelters project far from constituting an isolated transport system is actually a part of the urban furniture and equipment system for the City of Buenos Aires, also designed by Diana Cabeza and Martín Wolfson. Constituting a subsystem to the Transfer Centres, the Metrobus project responds to the same conceptual, formal and technological essence of the whole urban furniture of the City of Buenos Aires.

Designed with a strong visual character that grants it an energetic – but still integrated – urban presence, it includes as much of structural components and metallic enclosures as well as particular elements, such as railings, flagstones, rubbing bumpers, graphics, colour and lighting, designed especially for this project, with an autoparts criteria which in its association and interaction confer a determinant urban scale, constituting a convincing and simple system.

Thus in this project it can be said that the design covers all the project scales departing from the scale of the object, the scale of the "architecture" and the scale of urban landscape. Every part of this shelter system had a special design ad hoc and had a design with a serial production concept.

There are six existing typologies or functional bus shelters units that are to be accessed via a pedestrian ramp, given that they are elevated above the ground in order to grant a further direct accessibility for people to the vehicular units.

Its modular repetitive attribute allows to solve the needs of vehicule-passanger interaction along the whole of its length.

The bus stops are organised into structural modules of 3.3m, held 40cm above the height of the sidewalk.

The sequence proposes the following modules: with railing, with a tall backless seat, with average seats and with turnstiles.

Cenefas y paneles frontales

MIV / 02

MetroBUS
J. B. Justo

Buenos Aires
Gobierno de la Ciudad

Cenefas transiluminadas, paneles de orientación y de ubicación y paneles vidriados constituyen los elementos que funcionan como soporte de la información e identidad gráfica del sistema.

Referencias

Cenefa frontal

① a Pacífico →   MetroBUS Parada 14 : San Martín   a Pacífico →
Cenefa transiluminada de nomenclatura y orientación

Paneles frontales

② San Martín   a Pacífico →   Panel de orientación   ⑤   MetroBUS →   Panel de ingreso
③ 110   Panel de línea/recorrido   ⑥   MetroBUS   Panel Metrobus   Tablero eléctrico   ⚡
④   Panel topológico   ⑦ Parada 14 : San Martín   Panel de parada

D - 14. San Martín
Vista frontal | Esc 1:100

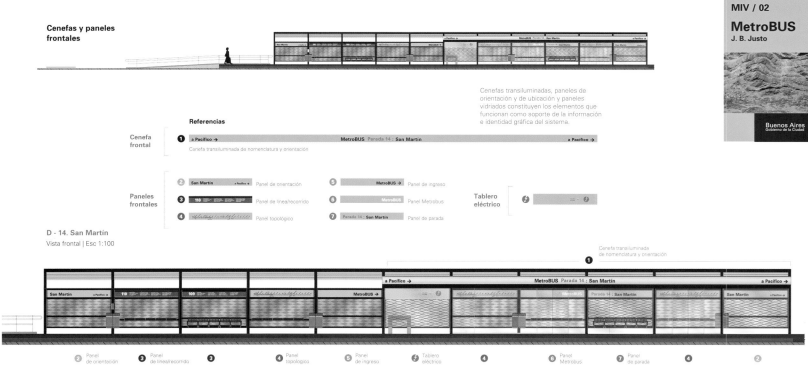

Cenefa transiluminada de nomenclatura y orientación

② Panel de orientación   ③ Panel de línea/recorrido   ③   ④ Panel topológico   ⑤ Panel de ingreso   ⚡ Tablero eléctrico   ④   ⑥ Panel Metrobus   ⑦ Panel de parada   ④   ②

　　都城巴士（快速公交系统）候车亭项目远非建立一个独立的运输系统。事实上，该项目是布宜诺斯艾利斯市城市景观与设备系统的一部分，也是由戴安娜·卡维萨与马丁·沃尔夫森设计。都城巴士项目是换乘枢纽子系统的组成部分，无论是在理念、形式还是技术层面上，都吸收了布宜诺斯艾利斯市所有城市设施的精髓。

　　该项目视觉性极强，赋予城市活力却又与城市面貌和谐统一。包括结构构件、导体机箱以及栏杆、石板、摩擦缓冲器、图案、颜色与灯光等特定元素，均为该项目特别设计。汽车配件标准相互联合、相互作用，形成了限定性的城市规模，该标准与以上元素共同构筑了一个简单有力的系统。

Panel frontal 4

Panel topológico

GRÁFICA EN FORMATO DIGITAL
02-Paneles frontales al

310 cm

D - 14. San Martín
Vista frontal | Esc 1:10

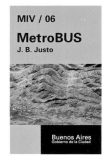

MIV / 06

MetroBUS
J. B. Justo

Buenos Aires
Gobierno de la Ciudad

El panel frontal topológico funciona como soporte de comunicación visual para la vista peatonal. Contiene:
- las estaciones que comprende el recorrido Metrobus.
- las líneas de colectivos que comparten tramos con Metrobus y sus destinos.

Producción gráfica y materialidad
Vinilo autoadhesivo sustrato blanco impreso, pegado sobre cristal 6mm + 6mm.

Fondo	Texto 1	Azul	Rojo	Verde	Naranja	Negro
Pantone 143 U	Pantone 412 U	Pantone 287 U	Pantone 032 U	Pantone 356 U	Pantone 166 U	Pantone Hexachrome Black U
C9 / M37 / Y95 / K0	C0 / M30 / Y66 / K98	C100 / M68 / Y0 / K12	C0 / M90 / Y86 / K0	C95 / M0 / Y100 / K27	C0 / M64 / Y100 / K0	C1 / M1 / Y1 / K100

NOTA 1: no se han calculado demasías para impresión. Se definirán según la tecnología y el proveedor en la etapa de implementación.

Tipo D | Mixta cerrada

7. Segurola / **14. San Martín**

Módulos 3 y 9

Cr	M	Cr
3		4

Asientos altos
(x5)

Nota: Se ejemplifica con la información del módulo 3

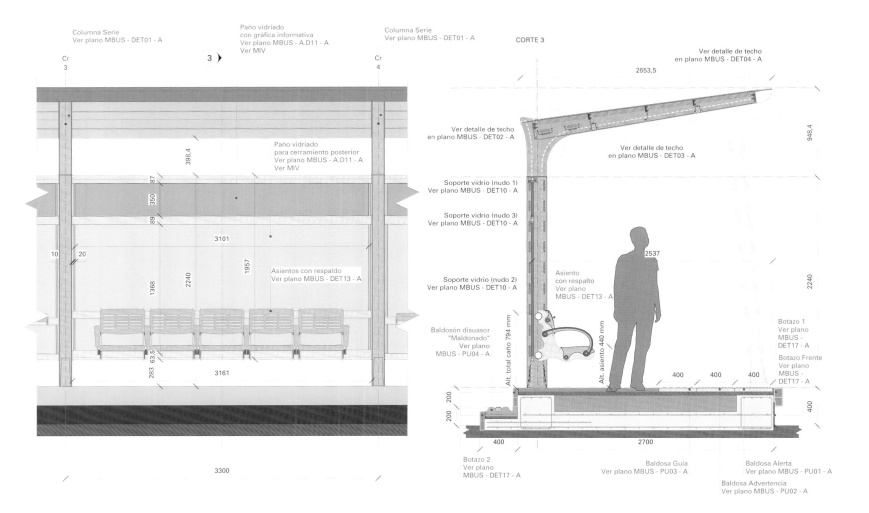

因此，可以说，在该项目中，设计包含了从物体规模，到"建筑"规模，再到城市景观规模的所有项目范畴。该候车亭系统每部分的设计都匠心独运，部件采用批量生产。

目前有六种类型或功能性候车亭。考虑到其位置高于地面，为了保证上车更加快捷，乘客穿过人行坡道，即可到达候车亭。

该项目不断运行，井然有序，解决了整个候车区内的人车交互问题。

巴士站位于高 3.3 米的结构模块中，比人行道高出 40 厘米。

接下来计划安装的模块有：栏杆、高高的无靠背座椅、普通座椅以及验票闸门。

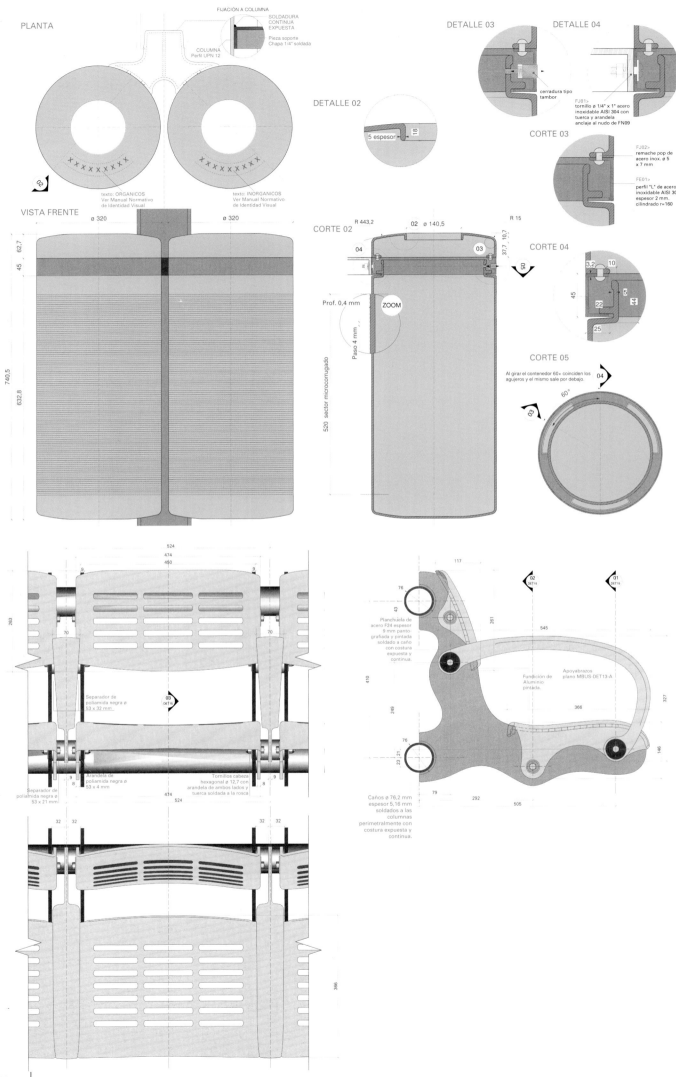

PLANTA

FIJACIÓN A COLUMNA

SOLDADURA
CONTINUA
EXPUESTA

COLUMNA
Perfil UPN 12

Pieza soporte
Chapa 1/4" soldada

texto: ORGANICOS
Ver Manual Normativo
de Identidad Visual

texto: INORGANICOS
Ver Manual Normativo
de Identidad Visual

DETALLE 02

5 espesor

18

VISTA FRENTE

ø 320
ø 320

62,7
45
740,5
632,8

520 sector microcorrugado

DETALLE 03

DETALLE 04

cerradura tipo
tambor

FJ01>
tornillo ø 1/4" x 1" acero
inoxidable AISI 304 con
tuerca y arandela
anclaje al nudo de FN09

CORTE 03

FJ02>
remache pop de
acero inox. ø 5
x 7 mm

FE01>
perfil "L" de acero
inoxidable AISI 304
espesor 2 mm.
cilindrado r=160

CORTE 02

R 443,2
02 ø 140,5
R 15

04
03

37,7 10,7

Prof. 0,4 mm

ZOOM

Paso 4 mm

CORTE 04

3,2
10
45
22
25

CORTE 05

Al girar el contenedor 60° coinciden los
agujeros y el mismo sale por debajo.

60°

03
04

524
474
450
9
9

263
70
70

Separador de
poliamida negra ø
53 x 32 mm.

05
DET15

Arandela de
poliamida negra ø
53 x 4 mm

Tornillos cabeza
hexagonal ø 12,7 con
arandela de ambos lados y
tuerca soldada a la rosca

Separador de
poliamida negra ø
53 x 21 mm

9
8
9
8
474
524

32 32
32 32

366

117

02
DET15

01
DET15

76
43

261

Planchuela de
acero F24 espesor
9 mm panto-
grafiada y pintada
soldado a caño
con costura
expuesta y
continua.

545

410

Apoyabrazos
plano MBUS-DET13-A

Fundición de
Aluminio
pintada.

366

249

327

76

146

23 21

79

292

505

Caños ø 76,2 mm
espesor 5,16 mm
soldados a las
columnas
perimetralmente con
costura expuesta y
continua.

Detalles Techo
Frente

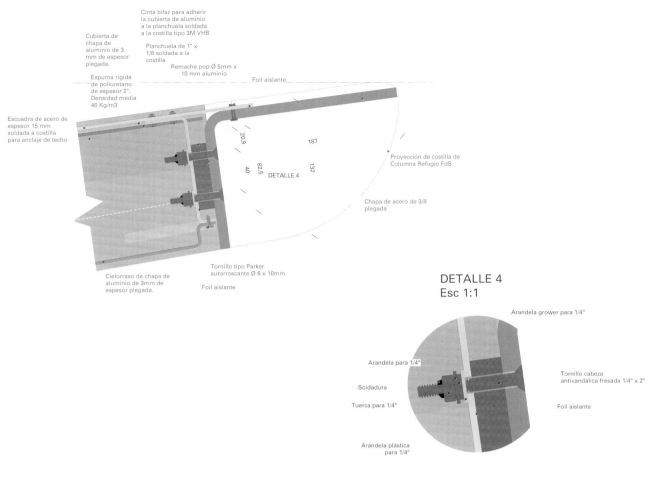

Cinta bifaz para adherir
la cubierta de aluminio
a la planchuela soldada
a la costilla tipo 3M VHB

Cubierta de
chapa de
aluminio de 3
mm de espesor
plegada.

Planchuela de 1" x
1/8 soldada a la
costilla

Remache pop Ø 5mm x
10 mm aluminio

Espuma rígida
de poliuretano
de espesor 2".
Densidad media
40 Kg/m3

Foil aislante

Escuadra de acero de
espesor 15 mm
soldada a costilla
para anclaje de techo

20,9

187

82,5

40

132

DETALLE 4

Proyección de costilla de
Columna Refugio FdS

Chapa de acero de 3/8
plegada

Cielorraso de chapa de
aluminio de 3mm de
espesor plegada.

Tornillo tipo Parker
autorroscante Ø 6 x 10mm

Foil aislante

DETALLE 4
Esc 1:1

Arandela grower para 1/4"

Arandela para 1/4"

Soldadura

Tuerca para 1/4"

Tornillo cabeza
antivandálica fresada 1/4" x 2"

Foil aislante

Arandela plástica
para 1/4"

Detalles Techo
Atras

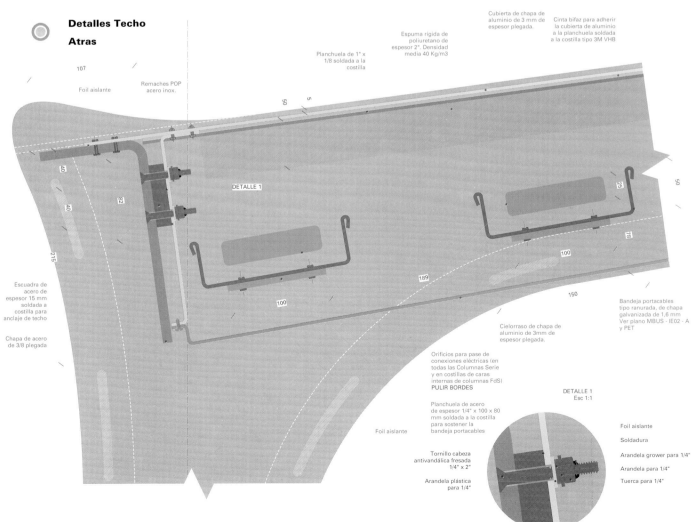

Cubierta de chapa de
aluminio de 3 mm de
espesor plegada.

Cinta bifaz para adherir
la cubierta de aluminio
a la planchuela soldada
a la costilla tipo 3M VHB

Espuma rígida de
poliuretano de
espesor 2". Densidad
media 40 Kg/m3

Planchuela de 1" x
1/8 soldada a la
costilla

107

Foil aislante

Remaches POP
acero inox.

50

5

40

40

215

32

DETALLE 1

50

50

78

100

189

150

100

Escuadra de
acero de
espesor 15 mm
soldada a
costilla para
anclaje de techo

Chapa de acero
de 3/8 plegada

Cielorraso de chapa de
aluminio de 3mm de
espesor plegada.

Bandeja portacables
tipo ranurada, de chapa
galvanizada de 1,6 mm
Ver plano MBUS - IE02 - A
y PET

Orificios para pase de
conexiones eléctricas (en
todas las Columnas Serie
y en costillas de caras
internas de columnas FdS)
PULIR BORDES

Planchuela de acero
de 1/4" x 100 x 80
mm soldada a la costilla
para sostener la
bandeja portacables

Foil aislante

Tornillo cabeza
antivandálica fresada
1/4" x 2"

Arandela plástica
para 1/4"

DETALLE 1
Esc 1:1

Foil aislante

Soldadura

Arandela grower para 1/4"

Arandela para 1/4"

Tuerca para 1/4"

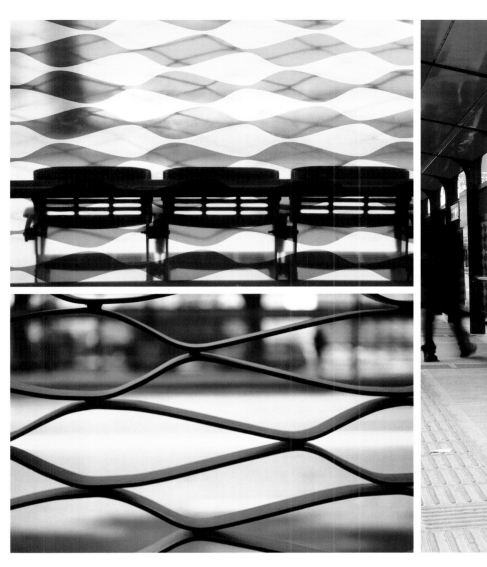

Limite metálico

Paño frontal

Medidas; detalles constructivos y unidad de la trama

Agujeros Ø 11 mm.

Distancias entre agujeros

Unidad de la trama

138,3

448,4

1515

1323

908

493

78

1401

54

3151

PROJECT DESIGNER
Diana Cabeza, Martín Wolfson, Leandro Heine

GRAPHIC DESIGNER
Osvaldo Amelio Ortiz, Gabriela Falgione, Pablo Cosgaya, Marcela Romero

Bus Shelter System

公共候车亭系统

Urban elements must respond to the geographical and cultural environments that generate them and be able to blend with their general disposition and particularities.

Bus shelters systems are intended to be placed on sidewalks; endowed with a front and a rear part, they allow people to walk throughout their perimeter and use them in different ways and from different directions. This creates continuity between the private realm of the bus stop and the open public space and confers a more dynamic perception of the city while avoiding the urban pathways that these elements usually generate.

The project focuses on preserving the historical heritage of the city, while providing modern urban elements for everyday life. Thus, they are designed in terms of a classical order, recovering the traditional urban industrial technology, whilst updating fabrication and maintenance procedures.

Total accessibility is present in the whole system of shelters and bus stops.

The system comprehends three different sizes of shelters and three different combinations of long row arrangements of each type of them. The basic bus stops, either self standing or attached to an urban wall, constitutes the minimum basic expression of the system.

The bus shelter system with the street signage system (see page 148) constitutes together the whole urban furniture for the City of Buenos Aires.

CLIENT
Government of the City of Buenos Aires

MATERIALS
Cast iron, painted steel structure, laminated green glass for bus shelters

LOCATION
Buenos Aires, Argentina

Bus shetler | C Type

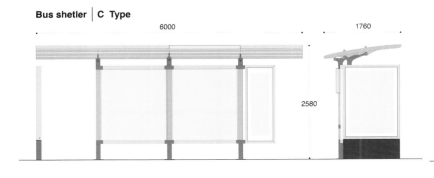

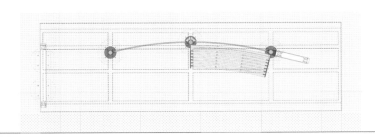

城市元素应当与孕育它的地理与文化环境相呼应，并且能够融入环境的整体布局和特质个性。

公共候车亭系统位于人行道旁，分为前、后两个部分，乘客可以穿行其中，采用不同的方式，不受方向的局限加以利用。这样一来就将公交车站的私人区域与公共开放空间衔接在一起，不仅避免了这些元素通常形成的城市道路拥堵，而且赋予了城市更多的活力。

这项工程在发展日常生活所需的现代城市元素的同时，保存该城市的历史遗产。

因此，候车亭是根据古典柱式结构设计而成的，不但再现了传统城市工业技术，而且体现出全新的工业制造和维修技术。

整个候车亭与公交车站系统为乘客提供了无障碍环境。

该系统由三个大小不同的候车亭构成，而每个类型的长排布置又有三种不同的组合。简单的公交车站、或独自屹立、或依附于城市墙壁上，构成了该系统最基本的样貌。

公交候车亭系统和标识系统共同构成了布宜诺斯艾利斯市的整套地景设施。

PASSAGEWAY

通道

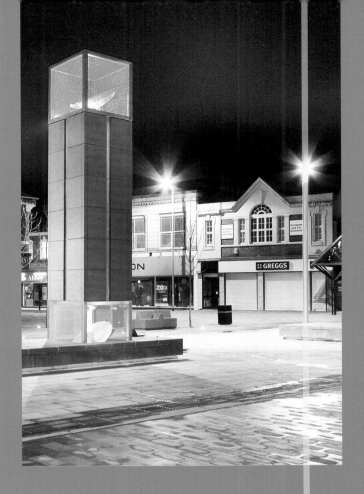

Covered Pedestrian Crossing
遮盖式人行通道

Desert Passage, Canopy Dreams
沙漠走廊，苍穹之梦

Pasarela del Postiguet
Pasarela Del Postiguet 环状步行桥

DESIGN COMPANY
Atelier 9.81

PROJECT TEAM
Lucie Vandenbunder, Geoffrey Galand + Cédric Michel, IOSISgroup, Les Produits de l'Épicerie

Covered Pedestrian Crossing

遮盖式人行通道

Downtown Tourcoing is currently at the heart of an extensive restructuring, launched a few years ago. All public spaces, streets and squares are being fully renovated, and a large shopping mall with movie theatres will be inaugurated soon.

As part of this project, the Metro, tram and bus station come together to offer a true multimodal system.

The project of a covered pedestrian crossing for downtown Tourcoing is born of this new direct relationship between transportation modes (with the bus station on one side and the trams and subway on the other), between the Place du Doctor Roux and the Place Charles et Albert Roussel.

The pedestrian crossing will fit into a row of townhouses of the same style, taking the place of one of them. By breaking thus with the alignment, the pedestrian crossing asserts itself visually, with the orange-red hues used on the open gables and by the constitution of a retro glassed façade, lit up at night.

Stepping into the void thus constituted, the project consists in erecting a canopy representing an urban origami.

Fine sheet of graphic Alucobond, a support for the signage designed with this project in mind.

Spread out over a 20-metre-long, 4.5-metre-wide area, this sheet reveals numerous complex folds and height variations, from which it derives its uniqueness.

The covering ends in a notable slope, signalling the pedestrian crossing from the Tram terminus and the entrance to the shopping mall and Metro.

The crossing's floor is made of grey granite pavement, an extension of the planned layout for all downtown public spaces.

CLIENT
SEM Ville Renouvelée

AREA
150 m²

LOCATION
Tourcoing, France

PHOTOGRAPHER
Julien Lanoo, Atelier 9.81

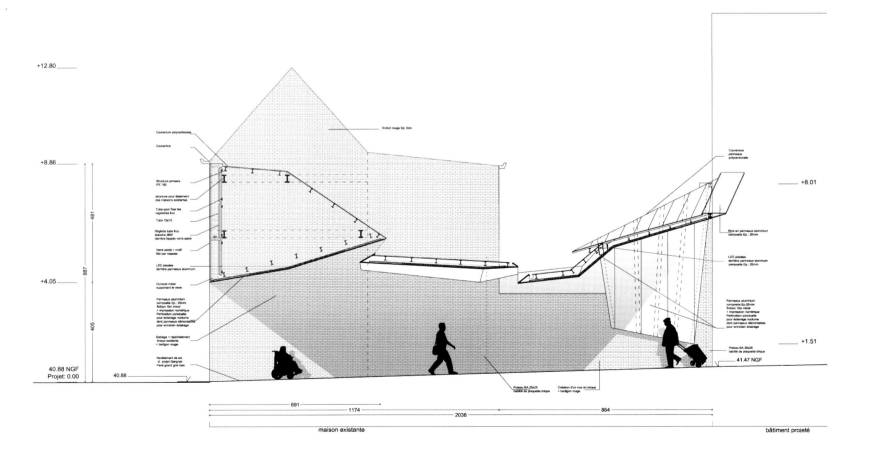

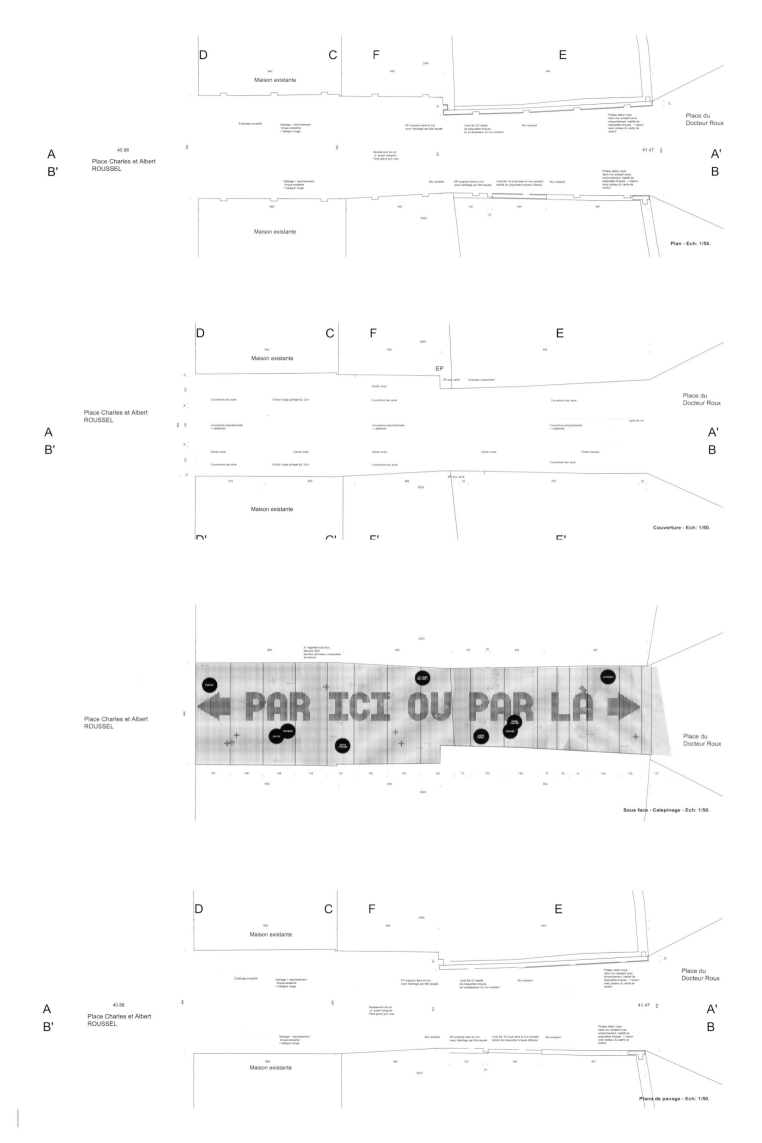

Plan - Ech: 1/50.

Couverture - Ech: 1/50.

Sous face - Calepinage - Ech: 1/50.

Plan de pavage - Ech: 1/50.

154

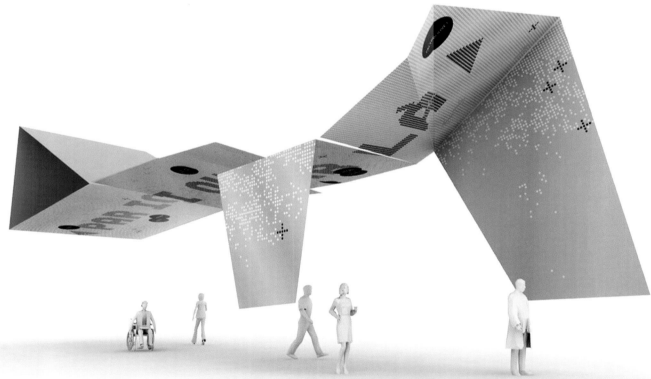

图尔宽市区当前是一个大规模重建的中心，该重建在几年前就开始了。所有的公共场所、街道和广场都在全面翻修，一个带电影院的大型购物商场不久将建成。作为这个项目的一部分，地铁、电车和公交车站会一起提供一个真正的多式联运系统。

图尔宽市区的遮盖式人行通道项目是这项运输方式之间的直接通道（公交车站在一侧，而电车和地铁在另一侧），在 Place du Doctor Roux 和 Place Charles et Albert Roussel 之间。

人行通道将融入一排排的同种风格的别墅，让这里成为它们的一部分。打破这样的工整，行人通道坚持了自身的视觉效果，在开口的三角形顶盖上运用橘红色色调，

在造型上使用复古的玻璃体立面，使它照亮了夜晚。

踏入这样构造的空间，该项目由一个建筑顶盖构成，呈现出一个折纸品形状。

好的 Alucobond 图形薄板，在脑海中是这个项目招牌设计的支架。

在一个 20 米长、4.5 米宽的区域铺开，这块薄板展现了许多复杂的褶皱和高度的变化，展现了其独特性。

顶盖以一个斜坡收尾，标志了从电车总站和其入口到购物商场和地铁的人行通道。

该通道的地板是由灰色的花岗岩铺成的，是所有市区公共场所的规划布局的延伸。

ARTIST	DESIGN COMPANY	DRAWINGS AND PLANS	DIMENSIONS
Barbara Grygutis	Barbara Grygutis Sculpture LLC	Barbara Grygutis Sculpture LLC	H 25' x L 200' x W 10'

Desert Passage, Canopy Dreams

沙漠走廊，苍穹之梦

Commissioned by Maricopa Community College District, Maricopa County, Arizona. An elevated sculptural pedestrian bridge connecting the new Ironwood Hall to an existing building on the Chandler Gilbert Community College Pecos Campus. This illuminated, curvilinear pathway meanders 200 feet through a grove of ironwood trees and highlights the natural beauty of the ironwood leaf. The laser-cut design in the hoops, natural daylight and integrated, designed lighting, create dramatic shade patterns along the pathway day and night. Barbara Grygutis collaborated with Architekton, the Architects for the New Ironwood Hall, to create an integrated, site-specific work of art.

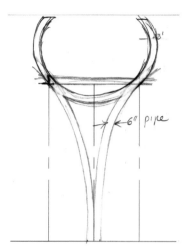

MATERIAL
Painted Steel, Laser-cut Aluminum, Designed Light

LOCATION
Arizona, USA

PHOTOGRAPHER
Barbara Grygutis, Margaret Kirkpatrick

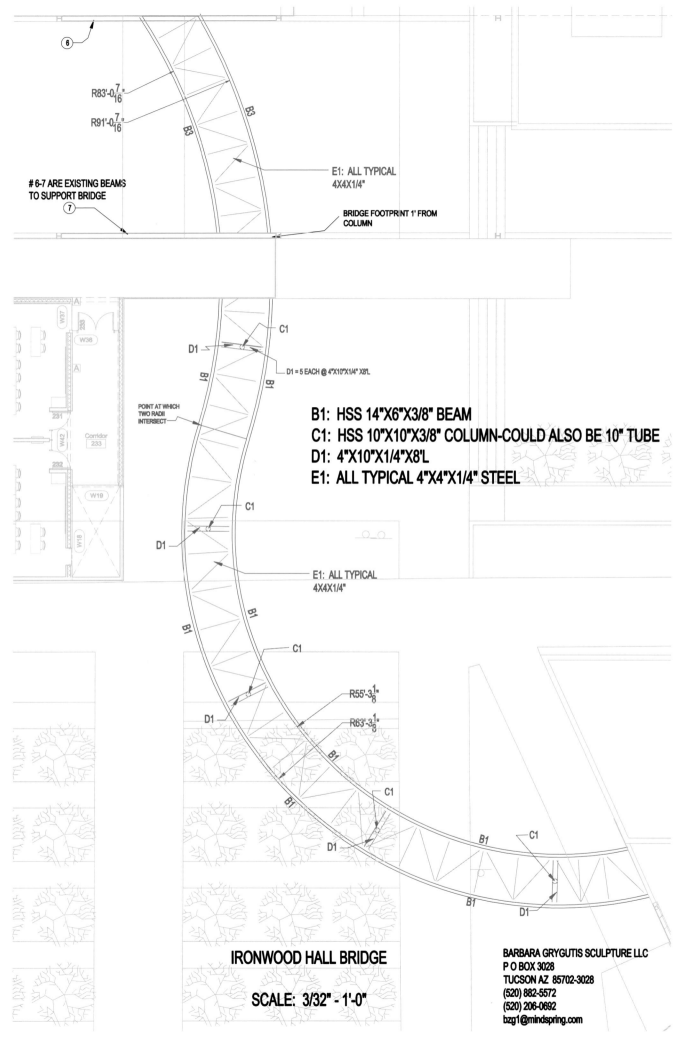

一个拔地而起的、雕塑般的人行天桥，连接了新的铁木大厅到钱德勒吉尔伯特社区学院佩科斯校园现有的建筑。这明亮的、曲折的通道弯转了 61 米，穿过铁木树林，渲染了铁木树叶的自然美。沿着通道的铁环栅栏的激光切割设计、自然光的聚集、照明的设计，为白天和黑夜创造了生动的光影图案。作为新铁木大厅的建筑师，Barbara Grygutis 与 Architekton 一起合作，创建了一个综合的、场地特定的艺术作品。

R83'-0 7/16"
R91'-0 7/16"

6-7 ARE EXISTING BEAMS TO SUPPORT BRIDGE

E1: ALL TYPICAL 4X4X1/4"

BRIDGE FOOTPRINT 1' FROM COLUMN

D1 = 5 EACH @ 4"X10"X1/4" X8'L

POINT AT WHICH TWO RADII INTERSECT

B1: HSS 14"X6"X3/8" BEAM
C1: HSS 10"X10"X3/8" COLUMN-COULD ALSO BE 10" TUBE
D1: 4"X10"X1/4"X8'L
E1: ALL TYPICAL 4"X4"X1/4" STEEL

E1: ALL TYPICAL 4X4X1/4"

R55'-3 1/8"
R63'-3 1/8"

IRONWOOD HALL BRIDGE

SCALE: 3/32" - 1'-0"

BARBARA GRYGUTIS SCULPTURE LLC
P O BOX 3028
TUCSON AZ 85702-3028
(520) 882-5572
(520) 206-0692
bzg1@mindspring.com

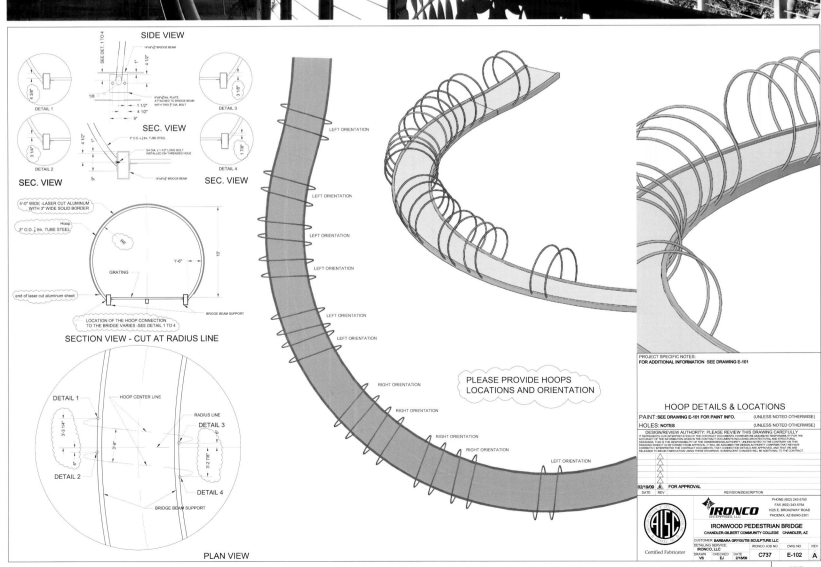

SIDE VIEW

DETAIL 1

DETAIL 2

SEC. VIEW

SEC. VIEW

DETAIL 3

DETAIL 4

SEC. VIEW

4'-0" WIDE -LASER CUT ALUMINUM
WITH 3" WIDE SOLID BORDER

Hoop
3" O.D. 1/4" thk. TUBE STEEL

GRATING

end of laser cut aluminum sheet

LOCATION OF THE HOOP CONNECTION
TO THE BRIDGE VARIES -SEE DETAIL 1 TO 4

BRIDGE BEAM SUPPORT

SECTION VIEW - CUT AT RADIUS LINE

DETAIL 1

HOOP CENTER LINE

RADIUS LINE

DETAIL 3

DETAIL 2

DETAIL 4

BRIDGE BEAM SUPPORT

PLAN VIEW

LEFT ORIENTATION

LEFT ORIENTATION

LEFT ORIENTATION

LEFT ORIENTATION

LEFT ORIENTATION

LEFT ORIENTATION

RIGHT ORIENTATION

RIGHT ORIENTATION

RIGHT ORIENTATION

LEFT ORIENTATION

PLEASE PROVIDE HOOPS
LOCATIONS AND ORIENTATION

PROJECT SPECIFIC NOTES:
FOR ADDITIONAL INFORMATION SEE DRAWING E-101

HOOP DETAILS & LOCATIONS

PAINT: SEE DRAWING E-101 FOR PAINT INFO.   (UNLESS NOTED OTHERWISE)
HOLES: NOTES   (UNLESS NOTED OTHERWISE)

DESIGN/REVIEW AUTHORITY: PLEASE REVIEW THIS DRAWING CAREFULLY

02/19/09   FOR APPROVAL
DATE   REV   REVISION/DESCRIPTION

IRONCO
ENTERPRISES, LLC

PHONE (602) 243-5750
FAX (602) 243-5764
1025 E. BROADWAY ROAD
PHOENIX, AZ 85040-2301

IRONWOOD PEDESTRIAN BRIDGE
CHANDLER-GILBERT COMMUNITY COLLEGE   CHANDLER, AZ

CUSTOMER: BARBARA GRYGUTIS SCULPTURE LLC

DETAILING SERVICE:
IRONCO, LLC

IRONCO JOB NO   DWG NO   REV

Certified Fabricator

DRAWN
VS   CHECKED
EJ   DATE
2/18/09   C737   E-102   A

159

DESIGNER
Susana Iñarra, Marin Marinovic

DESIGN COMPANY
BG STUDIO

PROJECT TEAM
Arch. Alex Ayala, Arch. Nenad Katic, Arch. Mónica Espí, Engr. Juan Moreno

Pasarela del Postiguet

Pasarela Del Postiguet 环状步行桥

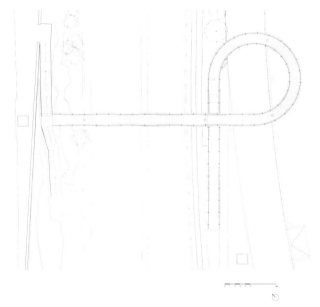

Alicante is a city that grew overlooking the Mediterranean Sea.

"Pasarela del Postiguet" is an essential element that connects its urban centre with the sea, spanning the Juan Bautista Lafora avenue which, with its intense traffic, presents a true visual and physical barrier between the beach promenade and the city. In this way Pasarela becomes an important link in the life of people of Alicante.

BG STUDIO's intervention consists in adaptation of a 20-year-old structure, on one side making it more appropriate to present technical and functional requirements, and on the other side redefining its aesthetic language into a symbol of the values that city of Alicante aspires to transmit: elegance and modernity.

Sculptural form of the new Pasarela del Postiguet was inspired by lightness of fresh mediterranean breeze, smoothness of the spiraling movement with which Pasarela descends from the hillside to the beach, and the dramatic forces that drive sailing boats at the horizon.

The existing metal structure was first completely restored and protected and then wrapped into a new, continuously curving, skin which changes its section along the entire length of the pasarela.

The skin consists of modules of plastic reinforced with resin and glass fibre (FRP). This material not only creates an astounding play of lights and reflection, but also has numerous advantages for this type of application, including very low weight, easy installation and its good resistance to mechanical impact and corrosion. Linear LED lighting is laid along the interior face of the modules which endows this sculptural object with a unique and dynamic identity.

LOCATION
Alicante, Spain

CLIENT
Ayuntamiento de Alicante / City Council of Alicante

COMPOSITES FACTORY
Talleres Xuquer SL

PHOTOGRAPHER
Javier Guijarro Tortosa

阿利坎特是一座远眺地中海的城市。

Pasarela del Postiguet 环状步行桥是一处基本的城市建筑，它跨越 Juan Bautista Lafora 大街——街上车辆川流不息，在视觉上形成海滨风光与城市建筑群之间的天然屏障，连通了市中心区和海洋。因此在当地居民的日常生活中 Pasarela del Postiguet 环状步行桥是一条重要的通道。

BG 工作室的重建计划包括对一处有 20 年历史的古老建筑进行改造，一方面要使它更加满足设计上和功能上的需要，另一方面要重新定义其艺术内涵，使其象征阿利坎特所希望传递的价值观——优雅与现代性。

Pasarela del Postiguet 环状步行桥的建造灵感，来源于轻快的地中海微风、

Pasarela 市由海滩盘旋延展至山丘的地貌特征以及推动帆船驶往地平线的巨大力量。

原先的金属构造首先被全部地修缮和保护，接着被弯曲成连续不断的新的弧度，部分构造发生了改变，整座步行桥沿着 Pasarela 市建造。

建筑表面被塑料模块覆盖，使用树脂和玻璃纤维加固（FRP）。这种塑材的使用不仅惊人地增强了光线和它的反射，还拥有其他诸多好处，比如轻便易装、抗腐蚀以及抗撞击。LED 灯沿着塑料模块内部直线设计，为这项建筑增加了一种独特、富有动感的特征。

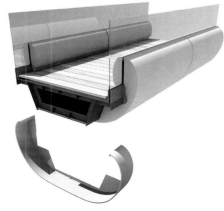

secuencia 01

secuencia 02

secuencia 03

PUBLIC SPACE

公共空间

LEAD PUBLIC ARTIST
David B. Dahlquist, Don Scandrett

DESIGN COMPANY
RDG Dahlquist Art Studio/RDG Planning & Design Architects, Des Moines, Iowa

OWER
The City of Omaha, Nebraska

Tree of Life

生命之树

The South Omaha "Tree of Life" streetscape installation embodies the concept of "cross-culturalism" and acts as a visual and symbolic gateway to South 24th Street. Five blocks have been transformed to create a dynamic multi-cultural experience and revitalised historic district in South Omaha, Nebraska. The project developed from a close cultural analysis of the South Omaha community that sought input and involvement from families, students and the local business association. The result is an aesthetic that celebrates cross-cultural folk art traditions of the major ethnic groups (Czech, Polish, Croatian, and Latino).Colourful textile patterns are interpreted in custom-glazed ceramic mosaic tile, pattern-cut metal, and lighting. A character and continuity is achieved through an integration of elements that feature benches, planters, and other sculptural features, meaningful and responsive to the concerns of the local community. The Tree of Life sculpture, over 36' tall, made of steel, with acrylic and state-of-the-art LED lighting, has become a new destination icon and cultural landmark.

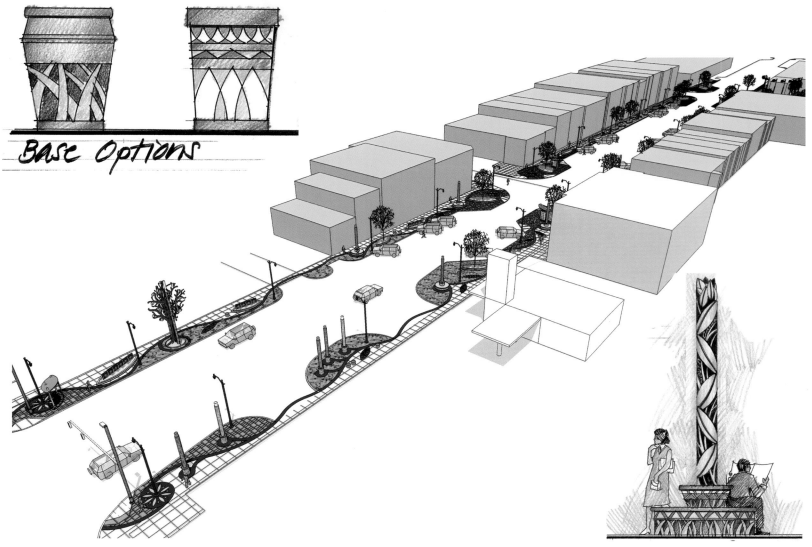

Base Options

TEAM
Travis Rice, Dolores Silkworth, Jonathan Martin, Pat Dunn Landscape Architects; Marty Shukert, Planner; David Raver, Lighting; Artist Staff, fabrication

MATERIAL
Iowa Metal Fabrication

GFRC
Artisan Stone

LOCATION
South Omaha, Nebraska, USA

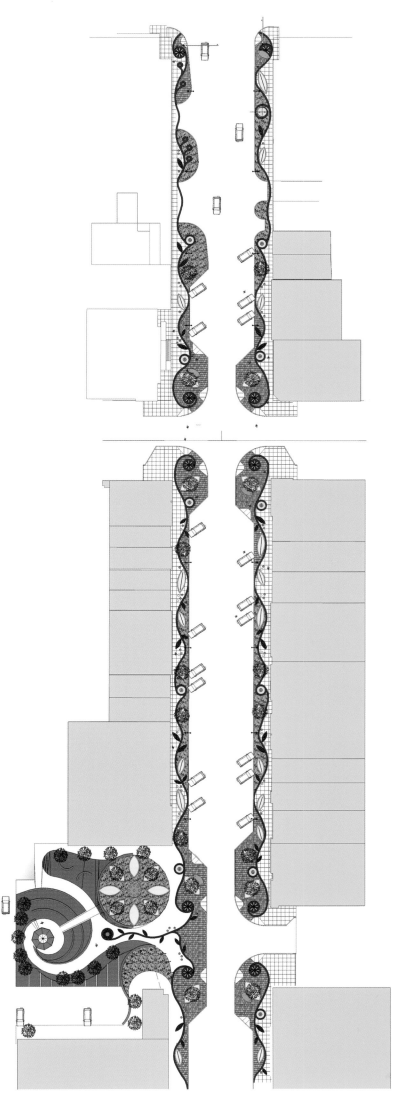

　　"生命之树"位于南奥马哈市，是一个著名的街景设施。它将"跨文化主义"的概念具体化，并且无论从视觉上还是象征意义上，它都是通往南第24大街的入口。为了创造一种动态的多种文化体验，振兴内布拉斯加州南奥马哈市的历史保护街，项目对周边的五个街区进行了彻底改造。设计师对南奥马哈市的文化进行了透彻的分析，从许多家庭、学生和当地商业协会那里搜寻各种相关信息。其结果就是形成一个审美原则，即歌颂几个主要民族的跨文化民间艺术传统（捷克、波兰、克罗地亚和拉丁美洲）。彩色的纺织模式可以通过定制的光滑陶瓷锦砖、切割金属和灯光来体现。通过各种元素的整合实现了项目的特点和连续性，包括独特的长椅、花架和其他雕塑般的装饰，它们既有意义，又满足了当地社区的需要。生命之树是一个高达10米的钢制雕塑，配有亚克力和最先进的LED照明灯，因此成为了新的旅游景点和文化标志。

21' 6"

19' 6"

35' 0"

29' 0"

20' 0"

6' 6"

1' 6"

1' 0"

DIA 9'

DESIGNER Alan Leo Pleština, Anamaria Filipović, Ivana Marić, Zrinka Babić, Krešo Ceraj, Sandra Brajković, Marko Duran, Berislav Medić, Goran Janjuš, Heinrich Gottwein, Darko Makar, Ivan Pešo

DESIGN COMPANY
UPI-2M

Landscape Design for Arena Centre

Arena Centar 景观设计

The studio UPI-2M was engaged to design a new modern and unconventional commercial complex within the suburban city area, which, according to new master-plan, is becoming the new centre of the city – the complex is situated next to the recently built multipurpose hall Arena Zagreb which is connected with the complex by three pedestrian ramps and a brand new public square. The complex and the hall generated a significant contribution to urbanity and development of the neglected and undeveloped city area. All this was a very challenging, but at the same time, delivered a strong responsibility for architects to design a creative, attractive and functional space around the complex. A different approach in working out this project conceived the idea of having a modern and unique outdoor landscape, instead of very usual environment that shopping centres have: a common view on a grand parking lot. A well arranged and designed landscape made a great contribution in originality, functionality and came as an added value of the whole area. The general idea is to present the variety of outdoor space in all segments. The landscape design embodies many different elements and is truly multifunctional. A newly formed public square with a public pedestrian cross-road is connecting three public pedestrian ramps coming from the multipurpose hall with the shopping complex and residential part of city area.

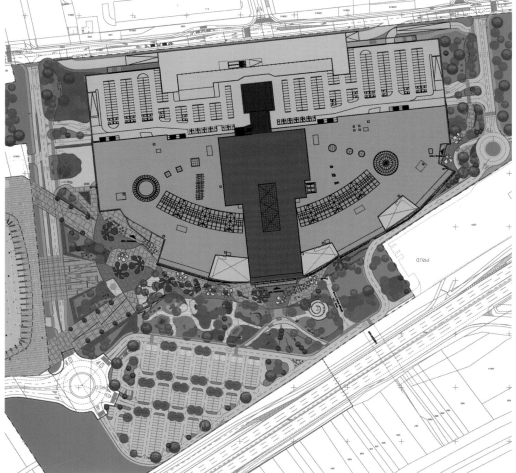

　　UPI-2M 工作室从事在郊区市区设计一个新的现代而非传统的商业综合大楼，根据新的总体规划，该郊区市区正在成为新的市中心——该综合大楼毗邻最近建成的多功能大厅 Arena Zagreb 旁，Arena Zagreb 的三条人行斜坡和全新的公共广场连接着综合大楼。综合大楼和多功能大厅对被忽视而又不发达的城区的都市化与发展产生了重大的贡献。这一切都很具有挑战性，但同时，需要建筑师有足够的责任感来在综合大楼周围设计一个有创意的、吸引人的、具有功能性的空间。制定该项目的一种不同的方法，就是构思具有现代和独特的室外景观的理念，而不是很平常的购物中心都有的环境：对一个大型停车场的一个普通的视野。良好的景观布置和设计在创意上、功能上和作为一个附加值的整个区域上，作出了巨大贡献。总体思路是，在所有片区呈现丰富多彩的室外空间。景观设计体现了多种不同的元素，是真正的一景多用。新建成的公共广场搭配了公共人行交叉路的三条人行斜坡，它们分别连接着多功能大厅、购物综合大楼和城市住宅区。

LOCATION
Croatia

AREA
87410.00 m²

PHOTOGRAPHER
Vanja Solin, Davor Konjikusic, Studio Blagec

DESIGNER
Alexander Lotersztain

DESIGN COMPANY
Studio Derlot

MANUFACTURER
Street & Garden Furniture Co.

Flinders Collection

弗林德斯系列

The Flinders Collection was designed to conform to the architectural features and archetypes. Great attention was placed on the number of seating components, its modularity and the way the different elements create flow and rhythm within the constraints of the side walk. Durability and robustness were paramount, with the design being able to withstand both natural forces and vandalism. The collection features an array of seating options including illuminated seats, drinking fountains, bollards, tree guards and grates and smoke stations.

弗林德斯系列的设计符合原型的建筑特色。座位部件的数量得到了极大的关注，它的模块化和不同的元素置于人行
道上的方式，限制了人行道上人流的节奏。耐久性和稳定性而又能够承受自然力量破坏的设计是最重要的。这个系列的
特征和陈列的座位，包括照明座椅、饮水机、灯箱、树木保护栏、栅栏和烟台。

DESIGN COMPANY
Tonkin Liu

MATERIAL Pre-cast concrete ramps, steps, landings, seawall coping pieces, and retaining walls;
weathered oak benches; light fixtures & light programming; light masts; stainless steel bins

LOCATION
Dover, UK

Dover's Esplanade

多佛海滨大道

plan - dover esplanade   **tonkin liu**

Dover's Esplanade has been given a radical overhaul by Tonkin Liu. The project has been conceived as three artworks, Lifting Wave, Resting Wave, and Lighting Wave.

The Lifting Wave is a repeated formation of sculptural ramps and staircases made of pre-cast white concrete that rise and fall to connect the Esplanade to the lower shingle beach. The Lifting Wave combines ramps formed of miniature steps that create a light-catching textured surface, with layered steps. The gentle ramps' sinuous line allow access for all and brings dynamic forms to the beach. The Resting Wave is a sculptural retaining wall that runs the length of the Esplanade, providing seating spaces sheltered from the south-westerly wind and orientated towards the sun. The wall is constructed with a shifting system of pre-cast white concrete blocks cast in stacked timber moulds, which result in a textured surface evocative of the sedimentary strata

layers of Dover's White Cliffs. The surface is designed to create shadows, minimising glare and discouraging fly-posting and tagging. Within the wall's recessed bays are benches made of weathered oak, shot-blasted and bleached to the texture of ocean driftwood.

The Lighting Wave is a sculptural line of white columns with artwork that captures the light, bringing improved amenity lighting and programmed lighting sequences to the Esplanade. Along the length of the Esplanade the columns rise and fall, catching the light of the day as well as creating a lighting feature at night. The Lighting Wave combines large flood lights to illuminate the Lifting Wave, medium spot lights to accentuate the undulating Resting Wave, and mini spot lights to be reflected by the artwork at its top. The interactive low-energy LED lights have been specifically programmed to create a dynamic wave movement, bringing a sense of delight to the seafront.

Tonkin Liu 对多佛的海滨大道进行了一次大刀阔斧的革新。该项目已被构思为三件作品：起伏波、休憩波、照明波。

起伏波是一个重峦叠嶂构造的雕塑坡道和阶梯、由白色预制混凝土砌成，用升降的形式衔接了滨海大道和相对低势的卵石海滩。起伏波结合微型阶梯的斜坡状，建造了一个层叠状的能采光的纹理表面。平缓的坡道蜿蜒伸展，给海滩带来了勃勃生机。休憩波是一个巨型的挡土墙、环绕着滨海大道海岸线，提供了挡避西南风并朝向太阳的休息场所。墙面是用层积材模具中的白色预制混凝土块的叠交方法构建

而成的、从而产生了多佛白崖的沉积层的纹理表面。墙面的设计产生了光影，最大限度地减少了炫光和阻碍广告和标签的粘贴。

照明波是一根艺术白柱的线性雕塑，它能采光，并给海滨大道带来改进的照明设施和照明顺序的编排。沿着滨海大道的海岸线，灯柱有起有伏，白天采集光线，夜间给予照明。结合了大型泛光灯的照明波，照亮了起伏波，中型的聚光灯突出照明了层层叠叠的休憩波，小型聚光灯则在艺术白柱的顶部照明。互动式的节能 LED 灯已经被专门规划来创造一个充满活力的波浪运动，给海滨带来了无穷乐趣。

R 814590

0m   25m

0m   5m

0m   1m

type AA   type A   type A   type AA

one CNC cut results in two opposing mould faces   convex type A and concave type AA share rotational symmetry   convex type A   concave type AA

type A   type AA   type B   type BB   type C   type CC   type C   type AA   type B   type CC   type A   type BB   type A

three variant types share rotation symmetry and common joint details   different arrangements of units give flexibility in plan form geometry

0m   5m

tonkin liu

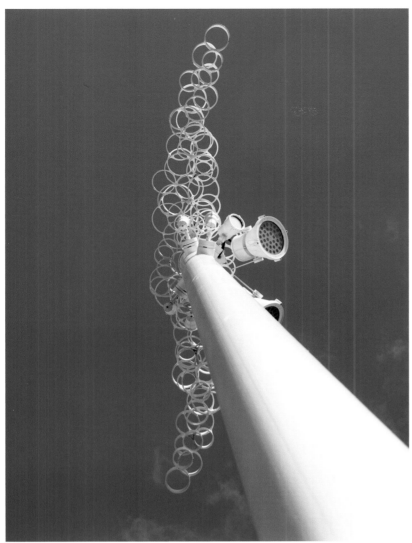

tonkin liu

lighting wave - dover esplanade

DESIGNER	DESIGN COMPANY	MANUFACTURER	PHOTOGRAPHER
Alexander Lotersztain	Studio Derlot	Street and Garden Furniture Co.	Florian Groehn

SIT Range

SIT 系列座椅

Southbank Institute of Technology public furniture

Under curator Jacquline Armistead, and in collaboration with Landscape Architects Gamble Mackinon Green, project Architects Cox Rayner and client South Bank Institute of Technology, designer Alexander Lotersztain set to design a comprehensive range of furniture for all outdoor areas of the site and a complimentary range of precast concrete benches named TWIG.

A collaboration with Manufacturer Street and Garden Furniture Company, the designer developed a range of 16 new products for the project. The products included an array of different seating configurations, tree grates, bollards, drinking fountains, workstations and bineclousures. The selection of finishes addresses the needs of this type of environment for both use and maintenance.

The brief accentuated the need for a strong durable and maintenance-free range. An 8mm steel galv. design was the designer's selection to undertake the concept development and design. A custom lime green anti-graffiti powder-coat was specifically developed for this range.

The colour selection and pattern with the designer's concept provides the relation between the range and Tafe's message.

"The pixel pattern representing the different areas that form the Tafe and the picture it creates when together; the colour promoting the idea of vibrancy and new vision 'the new heritage green'," said the designer.

南岸理工学院的科技公共家具机构

　　在馆长 Jacquline Armistead 的帮助下，加上与景观设计师 Gamble Mackinon Green、项目建筑师 Cox Rayner、南岸理工学院的顾客的通力合作，设计师 Alexander Lotersztain 开始为区域所有户外场所设计一个综合家居系列和

一个颇受赞誉的被称为 TWIG 的预制混凝土座椅系列。

　　制造商街区公司和花园家具公司联合开发，设计师为这个项目设计了一个拥有16 个新产品的系列。

　　产品还包括一系列的座位配置、树木护栅、短柱、饮水机、工作站和围栏。对

完成作品的挑选需要着重考虑到在这种环境下设施的使用和维护。

方案要以一系列耐用且免维护产品为重心。设计师选择 8 毫米的钢镀锌结构来表达发展与设计的概念。一种定制的绿黄色反涂鸦粉末涂层为这个系列特别设计。

有关设计师构想的颜色选择以及图案，将系列地和"Tafe"的信息联系一起。

设计师说："像素的造型代表了组成 Tafe 的不同区域，以及它们组合在一起的图像；颜色提升了思维的活力，展现出'绿色新传承'的景象"。

DESIGNER
Rehwaldt LA

DESIGN COMPANY
Rehwaldt Landschaftsarchitekten

LOCATION
Berlin Friedrichshain-Kreuzberg, Germany

PHOTOGRAPHER
Rehwaldt LA

Redesign Southern Lohmuehleninsel Berlin-Kreuzberg

南罗米伦岛的新规划，柏林－克罗伊茨贝格

The Island Square

The island theme was picked up as direct continuance of the playing strip. The slight topography provides new active opportunities. The surface is shaped with blue asphalt, partly covered with a protection layer (EPDM).

The slight ups and downs on the island square offer a reasonable space for playing around, skating and sliding. Within this sea of action the "Linseln" are sites of resting, recreation and observation.

Promenade

The path along the playing strip was slightly extended, so pedestrian and bicycle traffic could become separated. Bollards in the entrance area slow down cyclists at high speed. The surface was designed in large-size flagstone tiles which underlines the character of a boulevard.

Blue Moment

The transition zone along Goerlitz bank was redesigned and functional restructured. The site's sojourn quality is strengthened by unique seating furniture, called blue moment (Blaue Weile). It is a large, robust bench which stages the special look-out on the site. A special concrete frame is decorated by graffiti with the outlines of Goerlitz city. The bench is covered with wooden materials for comfortable seating.

Ark Stairs

The embankment is understood as an extended open space by a drawn-out staircase. A connection between the urban quarter and the channel emerged. From the stairs the view onto the flood channel, formerly called "free ark" (Freiarche), is staged while the quiet atmosphere on the waterfront can be experienced. On the sides, seating areas are arranged for additional sojourn opportunities.

Shipping Observation Plattform

Within the existing playground an additional playing element was designed - the "Schleusenwarte", a ship observation balcony. Right on the channel banks a balcony was implemented where passing ships can be welcomed. Therefore, metal elements which are installed on the balustrade can be moved upwards and downwards. By giving "flag signals", children will enjoy interacting with the passing ships.

岛屿广场

岛屿的构想建立在欢乐海岸不断延续的特点上。狭长的地形提供了新的机会来积极创造。建筑表面以蓝色沥青塑造，部分覆盖着一种保护层（三元乙丙橡胶）。

岛屿路面上有轻微的起伏，给溜冰和滑行的人提供了合适的游乐场地。 在这片生机勃勃的海域，"Linseln" 是休息、娱乐、观赏灯光的网络。

沿海长廊

沿着欢乐海岸细长蔓延的小路是那么适合步行，以至于就算脚踏车也要分开来骑。入口处的矮柱减缓了高速行驶的车速。长廊路面被设计为大面积的石板砖，这样更加突出了一种林荫小道的特点。

蓝色时刻

沿着 Goerlitz 银行的过渡区被重新设计，进行了着重于实用性的重建。通过被称为蓝色时刻（Blaue Weile）的独特座椅家居、区域的休闲功能被增强。座椅是一种体积大且坚固的椅子，它以一种醒目的外观使人们搜索到 Goerlitz 银行。特殊的混凝土框架装饰着绘有 Goerlitz 城市概貌的涂鸦。座椅覆盖着木质材料，使人们可以舒

适地坐下。

方舟路堤

路堤被理解为一个突起的台阶不断延伸出的开阔空间。一个市区间的连接和区域通道出现了。从楼梯上可以观看到原先被称为"自由方舟"（Freiarche）的涨潮海峡，这样的设计可以使人们体验到海滨的静谧氛围。另一方面，座椅区域的设置也为市民的额外逗留提供了可能。

轮船观赏平台

除了原本的游乐场地，还设计了一个新的游玩设施："Schleusenwarte"，一处轮船观赏平台。在狭道的右堤岸一处看台被设计，这里过往的轮船会很受欢迎。因此，在看台扶手上设置的金属元素上下移动。通过给以"标识信息"，孩子们可以享受与过往船只的互动。

DESIGN COMPANY
Uno+ Una. Arquitectura e Interiorismo Architecture

ARCHITECT
Plácido Piña,Esteban González, Rafael Calvent

LANDSCAPE
Noemí Zaro

Corredor Duarte

杜阿尔特走廊

The "Corredor Duarte" must be an opportunity to create an urban platform that benefits the city, creating a landscaping and lighting that celebrate structures, causing an urban spectacle for the benefit of pedestrian users, the countryside, urban and rural environment and the surrounding buildings.

The proposal pursues an ideal urban, landscape and coherent urban furniture in all the elevated, particularly caring for the individual characteristics of each location and the vocation of the city, to be governed by the latest trends of the current urban design for a more contemporary view of the whole.

It was decided to create a completely innovative built environment which aims to provide a daytime and evening attraction capable of inducing the public admiration, putting interest in economy, durability, maintenance and beauty of the structures.

Each element of the landscape and urban planning proposal was detailed, such as lighting and street furniture, vegetation and paving for better coordination with the execution of the work.

LIGHTING ARCHITECT ADVISOR
Gina Calventi

LOCATION
Santo Domingo, Dominican Republic

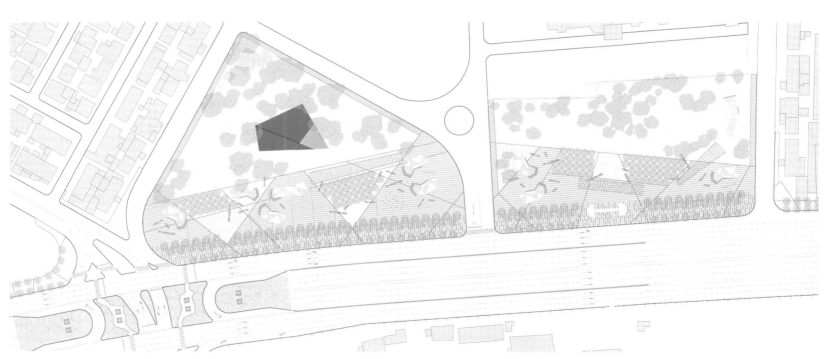

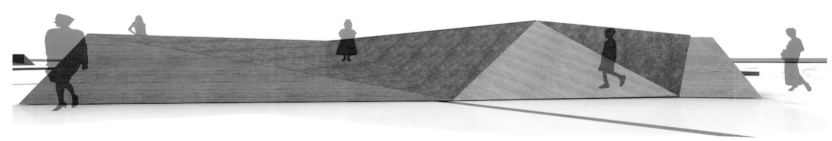

　　杜阿尔特走廊搭建起一个有益于城市发展的平台，创建了一个结构合理的环境美化工程和照明工程，形成了有利于行人、乡村、城市、郊区环境和周围建筑的城市景象。

　　该设计倡导的是高度理想的城市、景观和相关城市设施理念，同时要兼顾地方特色与城市功能，依照当下城市设计的最新趋势，使整体更具现代感。

　　经研究决定，要建立一个新颖的建筑环境，着眼于建筑的经济性、耐久性、维护与美观，使其在白天和夜晚都成为人们趋之若鹜的地方，获得公众的赞美。景观和城市规划设计的每一部分都很细致，例如灯光和街道装饰、绿化和铺石路面，这些都与施工完美结合。

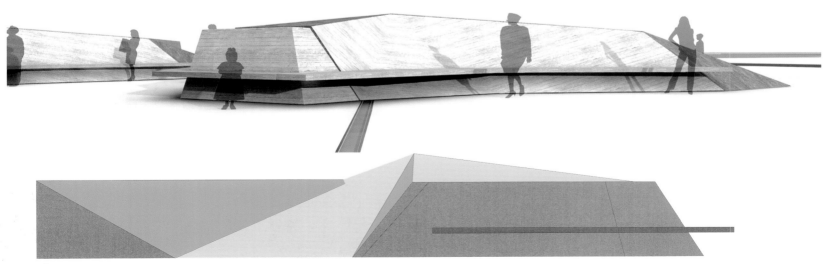

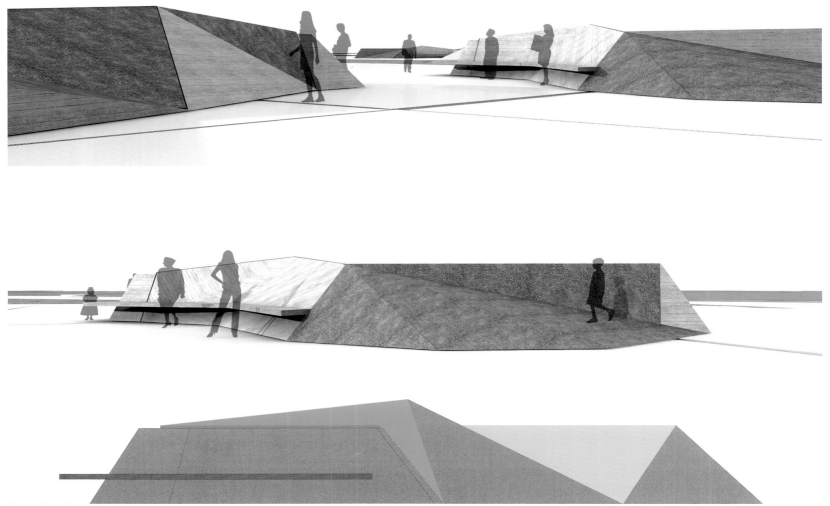

Dundalk Market Square

邓多克市集广场

BESPOKE MONOLITH: Height, 2,768mm, width 940mm, depth 300mm
S16 RECYCLING BIN: Height 1,172mm, width 1,160mm, depth 386mm, capacity 115L, weight 85Kg
S26 BOLLARD: Diameter/wall thickness 101/1.5mm, 114/3mm, 168/3.4mm; standard lengths 1,000mm, 1,200mm, 1,500mm
S71 CYCLE STAND: Height 875mm, width 950mm, outside diameter 48mm, wall thickness 2.77mm
S83 SEATING: Bench: length 2,000mm, width 540mm, height 460mm; Seats: length 2,000mm, width 540mm, height 777mm

Nicholas de Jong Associates (Urban Design) were commissioned by Dundalk Town Council to design the scheme. Following an extensive public consultation exercise, proposals were finalised and the contract to build was awarded to John Cradock Construction Ltd. (Co. Kildare). For Omos, this project represented an opportunity to engage with the Architect and Contractor at an early stage to ensure that his vision could be realised in an uncompromising yet economical way. Nicholas de Jong chose from Omos's standard range for much of the furniture as well as adapting some products to fit the project's specific needs. One of the key elements of the project is The Táin Sculptures by the Sculptor John Behan. The sculptures were originally commissioned by Mrs. Irene Quinn and displayed in the Imperial Hotel. In 1989 the sculptures were donated to Dundalk Town Council. They were subsequently attached to the exterior of the Tourism Office on the Market Square, where they remained until 2011. Based on the Architect's concept Omos produced eight illuminated cabinets just short of 3m tall to house the restored pieces. On the reverse of each cabinet etched on the glass are excerpts from The Táin telling the mythical story of Cúchulainn.

Bespoke monolith: Illuminated glazed display cabinets in powder-coated aluminium on a polished stone plinth. The Táin sculptures are framed to the front with exerpt to the rear.

s16 recycling bin: 3 and 4mm s355 steel, hot dipped galvanised throughout. Powder-coated finish on panels. Front opening with concealed stainless steel hinge and slam latch.

s26 bollard: Brushed 316 grade stainless steel with radially polished cap top section on painted galvanised steel lower section.

s71 cycle stand: 316 grade stainless steel or galvanised mild steel.

s83 seating: Cut granite ends with 316 grade stainless steel frame and treated iroko timber.

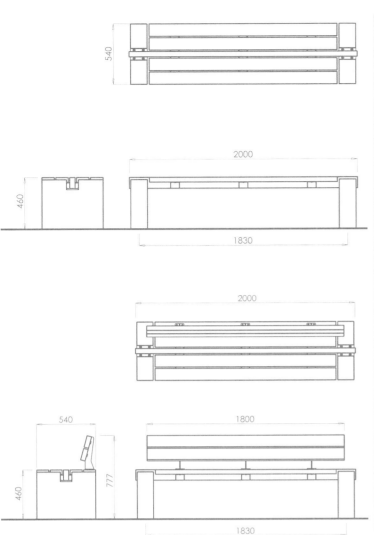

540

2000

460

1830

2000

540   1800

460   777

1830

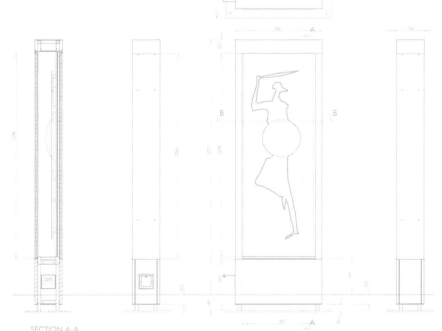

SECTION A-A

SECTION B-B

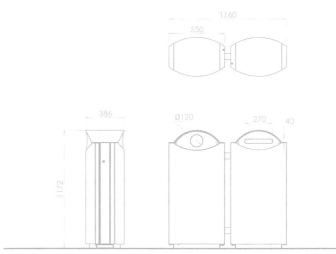

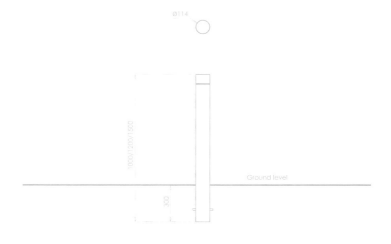

尼古拉斯·德容联合公司（城市设计）受邓多克市议会的委托，设计该项目。在开展最广泛的征询群众意见活动后，该建设意见书最终得以完成，同时，确定项目承建方为约翰克莱道克建筑有限公司（基尔代尔公司）。对 Omos 来说，这个项目提供了一个可以在初期与承包人和建筑师商讨的机会，从而确保他的想象能够以一种既保证质量又经济节约的方式实现。尼古拉斯·德容从 Omos 公司的标准范围中挑选了许多家具，同时为适应项目的特殊需求，改造了一些产品。雕塑家约翰·贝汉雕刻的 Táin 雕塑是该项目的关键要素之一。这些雕塑原本受到艾琳·奎因夫人的委托放置在帝国酒店内。1989 年，这些雕塑被赠给了邓多克市议会。随后，它们被放置在市集广场旅游部的外部，隶属于该部门，一直保留到 2011 年。基于建筑师的构想，Omos 制作了 8 个发光的陈列柜，高度将近 3 米，用来放置这些雕塑艺术品。每一

个柜子的背面都蚀刻了从 Táin 雕塑中选用的玻璃，讲述着库丘林的传奇故事。

s16 垃圾箱：3 毫米和 4 毫米的 s355 钢材，全部经过热浸镀锌。表面经过粉末涂层处理。前开口，内部由不锈钢转轴和锁舌构成。

定制的独石柱：发光的、上过釉的展示柜外面有一层铝制粉末涂层，放置在一块经过打磨的石头底座上。Táin 雕塑被框定在展示柜前面，后附摘录介绍。

s26 系缆柱：坐在经过油漆的镀锌钢底座上的为磨毛 316 不锈钢，顶部打磨得锃亮。

s71 脚踏车架：316 不锈钢或镀锌软钢。

s83 座椅：用切割花岗岩和 316 不锈钢框架以及经过处理的绿柄桑木制成

 STREET FURNITURE DESIGNER
Omos

CLIENT
Kildare County Council

LOCATION
Kilcock, Co. Kildare, Ireland

Kilcock

基尔库克

S11.3 LITTERBIN: Height 1,070mm, width 530mm, depth 400mm, capacity 90L, weight 62Kg

S16 LITTERBIN: Height 1,172mm, width 550mm, depth 386mm, capacity 115L, weight 85Kg

S56 BENCH: Length 2,000mm, width 615mm, height 455mm

S57 PLANTER (WITH SEATS): Overall height 745mm, overall width 1,600mm, overall depth 1,600mm, planter width 1,200mm, planter depth 1,200mm

S87 BOLLARD: Diameter 89mm, wall thickness 3mm, height 1,200/1,500mm

In November 2007 Kilcock, Co. Kildare underwent a significant on-street redevelopment programme. The work focused on the central part of the village and involved re-paving and kerbing, the introduction of new street lighting and street furniture.

Omos supplied all of the street furniture for the project. This included s11.3 litterbins, s16 litterbins, s87 bollards, s56 benches and s57 planters (some with bench seating, some without).

s11.3 litterbin: 316 grade stainless steel with satin finish outer construction on level adjustment galvanised base. Omos patented ashtray. Front opening with concealed stainless steel hinge and slam latch. 90L liner.

s16 litterbin: 3 and 4mm s355 steel hot dipped galvanised throughout, powder-coated finish on panels. Omos patented ashtray. Front opening with concealed stainless steel hinge and slam latch.

s56 bench: Cut stone ends and galvanised steel frame with provision for under-seat lighting. Treated hardwood surface with or without "anti-skate" bars.

s57 planter: 316 grade stainless steel on galvanised plinth. Iroko timber on two sides or around top to form bench.

s87 bollard: Painted galvanised steel with machined aluminium top.

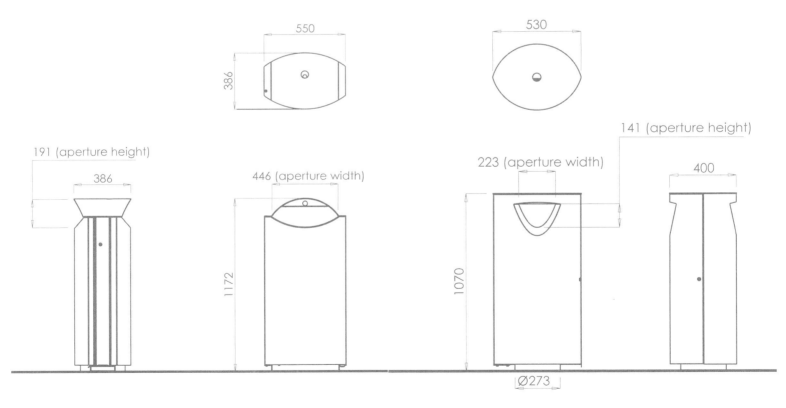

2007 年 11 月，基尔代尔郡的基尔库克开展了一个重大的街道重建项目。重点放在村庄的中心部分，涉及道路与边石的重新铺设，以及新街道照明与街道设施的引入。

Omos 为该项目提供了所有的街道设施，其中包括 s11.3 垃圾箱、s16 垃圾箱、s87 系缆桩、s56 长椅及 s57 花架（一些带有长条座椅，而另一些没有）。

s11.3 垃圾箱：水平调节的镀锌底座上是外部抛光的 316 不锈钢垃圾箱。拥有设计专利的 Omos 烟灰缸。前开口式设计，内部隐蔽着不锈钢转轴与锁舌。容量为 90 升。

s16 垃圾箱：整体采用 3 毫米与 4 毫米的 s355 钢热镀锌，上面经粉末涂层处理。拥有设计专利的 Omos 烟灰缸。前开口式设计，内部隐蔽着不锈钢转轴与锁舌。

s56 长椅：考虑到座椅下方的照明，将石头两端与镀锌钢架切掉。硬木表面经过处理，有的安装，有的则未安装防滑条。

s57 花架：镀锌底座上是 316 不锈钢，绿柄桑木材用于其两侧或顶部周围，形成了长椅。

s87 系缆桩：下面是刷了油漆的镀锌钢，上面为加工铝顶。

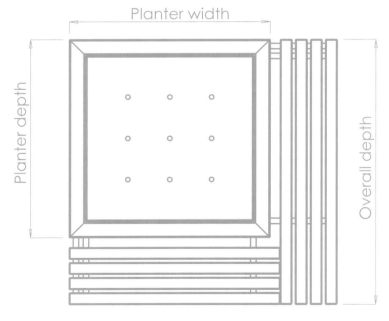

Planter width

Planter depth

Overall depth

12   Ø89

Overall width

Overall height

450

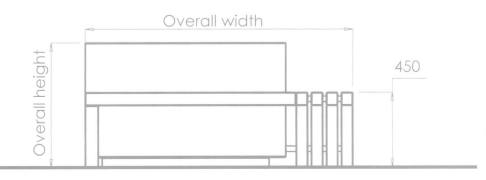

1200/1500

Ground level

300

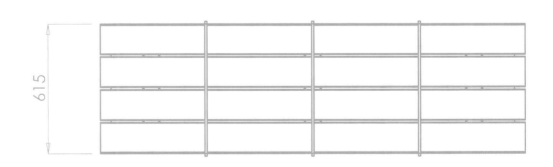

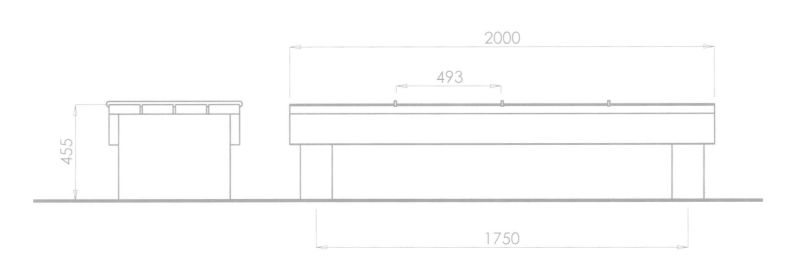

STREET FURNITURE DESIGNER
Omos

CLIENT
Carlow Town Council

LOCATION
Tullow Street, Carlow Town, Ireland

Tullow Street

图洛大街

S16 RECYCLING BIN: Height 1,172mm, width 1,160mm, depth 386mm, capacity 115L, weight 85Kg
T2 PLANTER: Height 970mm + 70mm adjustment, width/depth 1,000mm at base, 1,242mm at top
T3 SEAT: Length 2,000mm, depth 537mm, height 791mm

Tullow Street is the main shopping street in Carlow Town. The street was partially pedestrianised ten years ago when it was re-paved. The delineation between pedestrian and vehicular traffic was marked by colour and texture change across the level surface. Railings positioned intermittently along the street helped to mark change from road to footpath.

Since the street's last revamp it was often criticised for the lack of contrast between the street furniture, the road and the pavement. It posed particular difficulty for the visually impaired as the paving and furniture was said to appear as one single grey mass.

Carlow Town Council saw the opportunity to use brightly coloured street furniture to improve access to the street for the visually and physically impaired. Omos' "t" range was seen as a product which could achieve this goal. The simple yet bold design is intended to be finished in vibrant colours and be set against a grey backdrop so providing colour punctuation to the streetscape.

s16 recycling bin: 3 and 4mm s355 steel, hot dipped galvanised throughout. Powder-coated finish on panels. Front opening with concealed stainless steel hinge and slam latch.

t2 planter: Formed aluminium with powder-coated finish. Adjustable 316 grade stainless steel feet with brushed polish finish.

t3 seat: Formed aluminium body with powder-coated finish on brushed 316 grade stainless steel feet. Backrest in brushed 316 grade stainless steel.

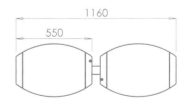

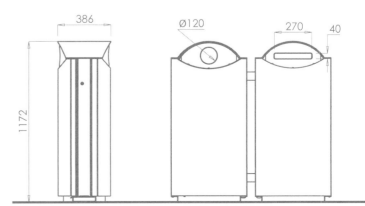

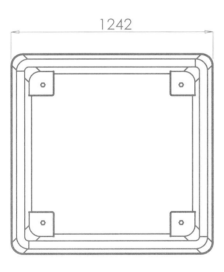

1242

970

1000

Adjustment + 70mm

图洛大街是卡罗镇主要的商业街。10 年前重新铺砌的时候这条大街部分是步行道。人行道和来往车辆的马路通过不同颜色和纹理在水平面标记出来。栅栏沿着街道等距放置，有助于区分马路和人行道。

从上一次整修之后，这条路就经常因为街道设施、马路和人行道之间缺少明显的对比而被诟病。人们觉得路面的铺设和道路设施就像一大块灰色的物体，对视觉障碍人士来说辨别困难。

卡罗镇议会会找准时机，用亮色的街道设施来改进街道入口，利于残障人士出行。欧姆斯的 t 系列产品能够达到这一目标。其设计简单却大胆，使用充满生机的颜色，与灰色的背景形成反差，为街道景观带来色彩亮点。

·s16 垃圾箱：3 毫米和 4 毫米 s355 钢材，全部热镀锌。表面经过粉末涂层处理。前开口、内部由不锈钢转轴和锁舌构成。

·t2 花架：经过粉末涂层处理的铝材。可调节 316 不锈钢底座，上面经拉丝抛光处理。

·t3 底座：经过粉末涂层处理的铝材坐在 316 不锈钢底座上。靠背使用表面拉丝处理的 316 不锈钢。

2000

520

792

430

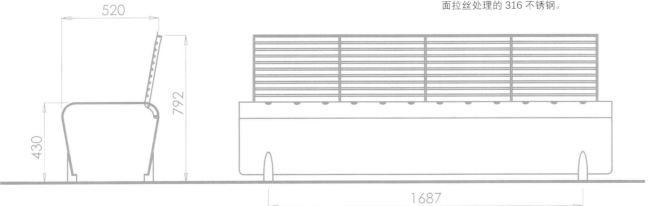

1687

DESIGN COMPANY	ARCHITECT		EQUIPMENT	CONSTRUCTION
 Microarquitectura	Martirià Figueras i Freixas, Paisajista. ASPECTE paisatge SLU.		Microarquitectura, S.L	Bruesa construcción, S.A

"Sistema U" in "Parc de la Corredora"

"Corredora 公园" 里的 "U 体系"

An intelligent use of urban facilities and structures that regenerates residual public space.

The Parc Corredora project activates a city area in disuse by retrieving an old irrigation channel "La Corredora", a natural watercourse. Its location is Castelldefels, a town near Barcelona. This intervention organises a residual area and transforms it in a new public space for recreational and sport uses. The entire project covers a surface of 21,624 square metres with a 700-metre walk.

It responds to the need to transform residual space in a place that citizens can use as a recreational spot, where sport can be practised and contact with nature is encouraged. Pedestrian passage, also suitable for cycling, is a linear path that adapts to the sequence of public spaces through which it passes.

To give unity to the project, an organising element is designed. It is a spatial structure with a rhythmic sequence formed by cut steel frames that enhances directionality. The function and size of these frames varies along the promenade. In areas where shady area is required, a framework supports recycled material panels (HPL leftover from making kiosks and recycled plastic panels coming from containers). These plates form an irregular mosaic of different colours, die-cutting elements, so the cast shadow is also irregular.

In cross section, frames height also varies depending on the area in which they are located, using these different heights to stand out some spaces.

Lighting is solved using frames themselves for locating fluorescent lamps, together with other lamps as independent columns of extruded aluminium.

The project is a good example of how a smart selection and application of urban facilities completely changes site character and promotes citizen interaction.

STRUCTURES	CLIENT	AREA	PHOTOGRAPHER
Joan Vilanova, Ingeniero	Ayuntamiento de Castelldefels	21,624 m²	Marc Guillen + Microarquitectura

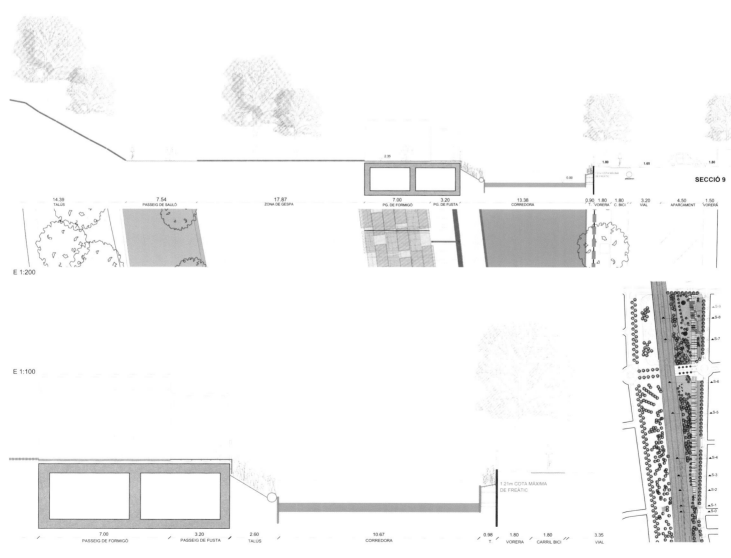

SECCIÓ 9

E 1:200

E 1:100

14.39	7.54	17.87	7.00	3.20	13.38	0.90	1.80	1.80	3.20	4.50	1.50
TALÚS	PASSEIG DE SAULÓ	ZONA DE GESPA	PG. DE FORMIGÓ	PG. DE FUSTA	CORREDORA	T.	VORERA	C. BICI	VIAL	APARCAMENT	VORERÁ

1.21m COTA MÀXIMA
DE FREÀTIC

7.00	3.20	2.60	10.67	0.98	1.80	1.80	3.35
PASSEIG DE FORMIGÓ	PASSEIG DE FUSTA	TALÚS	CORREDORA	T.	VORERA	CARRIL BICI	VIAL

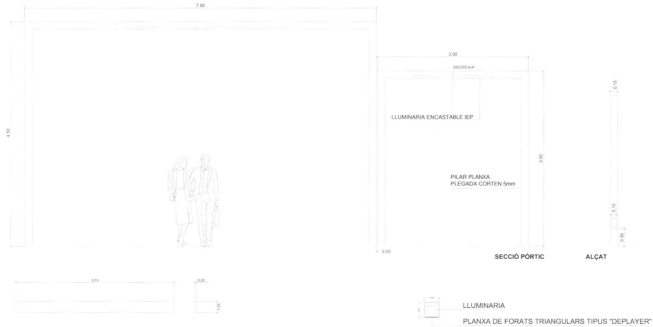

Detall lluminaria Mod FL-3 (330216) de IEP ILUMINACIÓN o similar

LLUMINARIA ENCASTABLE IEP

PILAR PLANXA
PLEGADA CORTEN 5mm

SECCIÓ PÒRTIC   ALÇAT

LLUMINARIA

PLANXA DE FORATS TRIANGULARS TIPUS "DEPLAYER"

SECCIÓ A-A'

合理使用城市设施、建筑，和再度利用剩余的公共空间。

通过重新开发古老的灌溉河道 "Corredora" ——一个天然河道，"Corredora 公园" 的项目激活了废弃的城区。它位于 Castelldefels，一个靠近巴塞罗那的小镇。相关部门将剩余区域变为新的供娱乐和锻炼用的公共场所。整个项目占地面积 21,624 平方米，有一条 700 米的走廊。

把剩余空间变成市民们可以使用的、亲近自然的、休闲与运动的地方，基层管理部门回应了这一需求。沿路的人行通道也同样适于骑自行车，那是一条笔直的小道，适合它经过的公共场合的顺序。

为使项目达到统一而设计了其组成要素。这是一个有节奏规律的排序、由切钢架组成，增强了方向性。这些框架的功能和大小沿着人行长道变化着。在需要阴凉的区域，有一个支持可再生材料面板的构架（制作凉亭的 HPL 剩料和来自集装箱的可再生材料面板）。这些板材形成颜色各异的不规则马赛克，利用了模切原理，所以投影也是不规则的。

在横截面上，框架的高度会有所不同，具体取决于其所在的区域，以不同的高度来突出一些空间。

通过使用荧光灯自身框架的定位，与其他作为挤压铝材独立柱的灯一起，使照明问题得到了解决。该项目是一个很好的例子，体现了一个非常聪明的选择，对城市设施的应用，能彻底改变场所的性质和促进公民的相互交流。

DESIGNER	DESIGN COMPANY	LOCATION	PHOTOGRAPHER
John Potter	Boffa Miskell	Auckland, New Zealand	Boffa Miskell

Te Wero Island and Eastern Viaduct

温耶德岛和东部高架桥

In 2010 Waterfront Auckland commissioned Boffa Miskell to design a five-year, temporary route-marking installation to emphasise the new pedestrian route that now crosses open space previously used for car parking.

Broad bands of boldly coloured surfacing mark out the new route, which leads from Quay Street through the Eastern Viaduct onto Te Wero Island. From there, the recently completed Wynyard Crossing pedestrian/cycle lifting bridge provides direct access to Karanga Plaza adjoining the newly opened Viaduct Events centre.

New road marking technology (Ballatini-Strada resin surfacing incorporating recycled coloured glass beads) enabled the use of bold and vibrant colour bands to transform the existing asphalt, concrete and clay paved surfaces into a visually striking and coherent route connecting Wynyard Quarter and Quay Street.

Movable tree planters containing semi-mature Pohutukawa trees and incorporating seating structures line the route and provide a physical separation to adjacent areas of car parking. At the threshold to the Wynyard Crossing lifting bridge brightly coloured movable chairs have been placed on an area of artificial lawn to create an outdoor lounge, where people can linger, relax and enjoy the unique view. Several converted shipping containers will also be placed adjacent to the lawn to provide accessible indoor shelter. It is intended that these structures are suitable for re-use when the installation needs to be relocated to cater for temporary events on site.

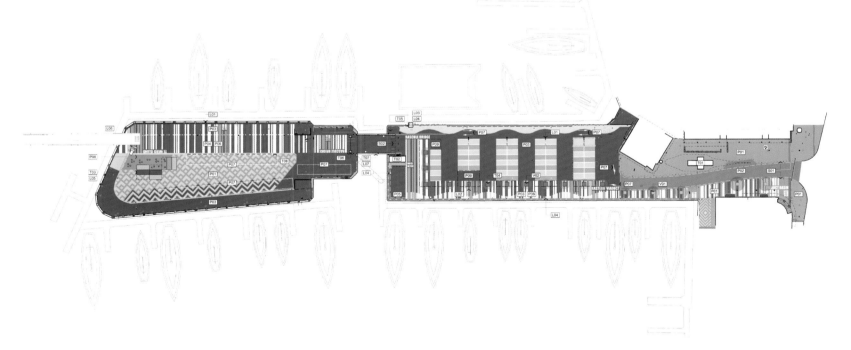

在 2010 年奥克兰海滨地区委托设计师 Boffa Miskell 设计一个五年的短期道路指示来标明新的步行路线，路线现在穿过以前被用来停车的开放区域。

醒目的彩色宽带标出了从奎伊街经过东部高架桥通往温耶得岛的新线路。从这里，最近竣工的温耶德步行十字路 / 循环升降桥提供了到 Karanga 广场的直接通道，广场靠近最近开放的高架桥活动中心。

新的道路标记技术（Ballatini-Strada 树脂表面与再生有色玻璃珠结合），使用大胆和鲜明的颜色带改现有的沥青混凝土和黏土表面，铺成连接温耶德区和奎伊街的一个引人注目的连贯路径。

种植圣诞树树苗的可移动的种植盆，以及与它合并的座椅设施沿着道路排成一行。在汽车停车处的临近区域提供了一个实体分离区。在温耶德交叉升降桥的入口处，色彩鲜艳的可移动座椅已经被放置在一个人工草坪区域创造一个室外的休息室，人们在那里可以徘徊、放松、享受独特的风光。几个改装的集装箱也将会安放在这个草坪旁边，用作室内庇护所。其目的是考虑到当场地发生临时事件需要重新布置，这些设施可以重新利用。

DESIGN COMPANY
scape landscape architects

TEAM
Anna Hardenberg, Miriam Kühn, Sebastian Riesop, Corinna Scheele, Andrea Tofall, Jochen Westhauser

Prinzenplatz

Prinzenplatz 广场

The "Prinzenplatz" square in the central gathering area in the town centre of Kamp-Lintfort. The extreme heterogeneity of the architectural fringes of the space needed to be counterbalanced with a strong design gesture in the centre. By reducing existing fixtures, shrubs and planted areas, a solitary island of four sycamores was created as the middle of the square, which is raised from the ground level by a broad step. The space is dominated by its generous stone surfaces.

The signature elements of the "Prinzenplatz" are the red bus stop shelters and curved benches - furniture that was custom-made for this place. The design is inspired by the story Kamp-Lintfort as a former mining site. Black stones

and red metal and wood are the materials used to create the elements. The curved benches define the centre of the square as a public lounge area of the residents. Nobody sitting alone on standard separated benches but all sit together in the middle of the city centre. Since the place has a lively nightlife, the bank is lit from inside with red light, so a warm, glowing impression arises. The very generously sized bus stop shelter follows in designing industrial manufacturing processes. The solid-looking roof resembles a bent piece of steel, which was established to stabilise on a massive concrete block. Numerous light fields appear in the waiting area, so a uniform pleasant light is created.

LIGHT PLANNING
licht · raum · stadt, Wuppertal

PROJECT MANAGEMENT
Matthias Funk

CLIENT
Stadt Kamp-Lintfort

LOCATION
Kamp-Lintfort, Germany

AREA
8,400 m²

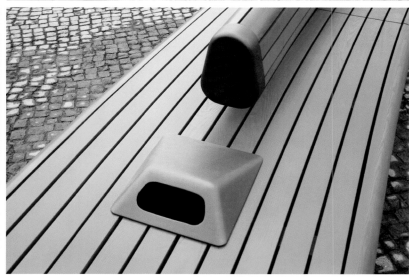

Prinzenplatz 广场是坎普林特福特市中心的主要聚集地点。广场的周围建筑具有很强的异质性，因此在中央需要一处具有强烈设计风格的景观来抵消周围的影响。通过减少原有的固定设施、灌木种植区域、建成拥有 4 棵悬林木的独立小岛来为广场的中心，小岛是一个从地面抬起的宽阔平台。

Prinzenplatz 广场的显著标识是红色的巴士候车亭和特有的弧形长椅。设计的灵感来源于坎普林特福特作为旧是采矿地点的故事。黑色的石头以及红色的金属被用来创造景观中的故事元素。

弧形的座椅使广场的中心成为一处市民休闲区域。市民们成群结队地坐在分离的标准座椅上，但是所有人都一起坐在市中心的广场中央。因为有热闹的夜生活，银行也从室内点亮红色的电灯，因此这里给人一种温暖、炽热的感觉。

宽大的巴士候车亭根据工业制造过程设计。坚固的亭顶外形类似一个弯曲的钢条，搭在一个巨大的混凝土石砖上。大面积的灯具设置在等候区，所以一种标准的令人愉悦的光照环境产生了。

DESIGN COMPANY	LOCATION	CLIENT	STRUCTURAL ENGINEER	AREA
SLA	Frederiksberg, Copenhagen	Municipality of Frederiksberg	Hansen & Henneberg	18,000 m²

Frederiksberg New Urban Spaces

腓特烈堡的新城市空间

The centre of Frederiksberg is the most densely populated area of Copenhagen. It lacked, however, quality public spaces for the more than 30,000 visitors that each day passes through the area. The Municipality of Frederiksberg turned to SLA to create new urban spaces bustling with life.

The Frederiksberg City Centre is a 20,000-square-metre public space dominated by large, introvert buildings. As a consequence, SLA reversed the relation between the inside and the outside. Instead of creating urban spaces that were subordinated to the buildings, SLA created five urban spaces that in themselves were "rooms" in their own right, using the surrounding buildings as façades. Using a method borrowed from soap bubbles, the five separate urban spaces were wedged intimately together, while at the same time providing each urban space with its own, distinctive feel and expression.

Moving through the different urban spaces, the visitors will experience radically different sensations: trees, fragrances, colours, surfaces, sounds (even the temperature and climate) will change from one urban space to the next. The only common denominator between the five urban spaces is that they are all extremely dependant on the weather, changing feel and expression with the rain, sun, and the wind. At nighttime the urban spaces change character with lights and artificial sound effects, creating individual experiences to be shared by all.

As a result, Frederiksberg City Centre has evolved from an introvert place for transit to a bustling urban space rich in expression and atmosphere. Now inhabitants and visitors come to the Frederiksberg City Centre to linger in the cafés, meet other people, or simply take a walk through the five different spaces, experiencing all the changing moods, sensations, and expressions life in a great city can offer.

腓特烈堡中心是哥本哈根人口最稠密的地区。然而，这个每天有超过 3 万人次的游客经过的地方缺乏优质的公共场所。腓特烈堡市政府委托 SLA 公司来设计一个可以热闹生活的新城市空间。

腓特烈堡市中心是一个 2 万平方米的的公共空间，由大型建筑和内向型建筑占主导地位。因此，SLA 公司扭转了内部和外部的关系。SLA 设计公司利用周围的建筑物作为表面，建造了五个其右侧都是"公共设施"的城市空间，而不是建立附属建筑的城市空间。利用从肥皂泡参考来的方法，使五个独立的城市空间紧密地联系在了一起，与此同时，让每个城市空间拥有其自身独特的感觉和表达。

经过不同的城市空间、游客将体验到完全不同的感觉：树木、芳香、颜色、外观、声音（甚至是温度和气候），从一个城市空间到另一个城市空间将有所变化。五个城市空间之间唯一的共同点是，它们都非常依赖天气、用雨滴、太阳和风转换感觉和表达。在夜间，城市空间通过灯光和人造音效来改变特征，创造了可以与众人分享的个人体验。

所以，腓特烈堡市中心已经从一个内向型空间过渡到一个富有表达和气氛的繁华的城市空间。现在，居民们和游客们来到腓特烈堡市中心，流连于咖啡吧、与他人邂逅、或者只是走过五个不同的空间散个步，体验所有心境上的、感觉上的、以及在一个伟大的城市提供的生活表达上的变化。

DESIGN COMPANY   MATERIAL

BURNS + NICE   Granite kerbs, concrete flags and granite paving in special areas, planting

LOCATION

London, UK

East London Transit Phase 1A

东伦敦运输道路 1A

The East London Transit (ELT) was designed as a fully integrated public transport system linking local town centres with areas of housing and employment. It connects to existing and future transport developments such as Crossrail, C2C railways, London Underground and Overground as well as local bus services.

Phase 1A was opened on 20 February 2010 and operates between Ilford Town Centre and Dagenham Dock via Barking Town Centre.

During the concept design stage a Strategic Urban Realm Plan (SURP) was developed. It placed the transit route in an 800m wide study corridor.

The findings of this study formed the basis for the design of the route corridor. Working in partnership with the affected London Boroughs and other stakeholders a holistic urban design solution was established that unifies the infrastructure needs for the transit route with wide ranging improvements to the public realm. An additional key aspect to the scheme was the development of an uninterrupted ELT branding. The route corridor has now become a continuous high quality public realm, highly legible and accessible to all users.

SIZE   PHOTOGRAPHER

10,000 m   BURNS + NICE, Transport for London, Tim Crocker

设计东伦敦运输（ELT）的目的是将其作为一个完全集成的公共交通系统，连接当地的镇中心、住房供给和就业领域。它结合了现有的以及未来的交通运输的发展，如横贯铁路、C2C 铁路、伦敦地铁和地上交通，以及当地的巴士服务。

1A 道路于 2010 年 2 月 20 日开放，在伊尔福镇中心和达格南码头经由巴金镇中心之间通车。

在概念设计的阶段，一个城市战略领域计划（SURP）已在开发阶段。它把交通路线设在一条 800 米宽的实验过道上。这项研究的成果是使道路通道的设计基础初步成形。与被涉及的伦敦城区和其他利益相关者合伙建立一个整体的城市设计方案，该方案需要把基础设施统一起来，大范围改善公共领域的运输路线。这个方案的另外一个重要方面是发展一个不间断的 ELT 品牌。这条道路通道现在已经成为一个持续的、高质量的公共领域，其具有高辨认度并对所有使用者开放。

Ilford Town Centre

Ilford Lane

Ilford Lane Shopping

Ilford Lane/Fanshawe Avenue Residential

Barking

Barking Town Centre

Barking Residential

Thames View

A13/Industrial

Bastable Avenue Residential

Dagenham Dock

Dagenham Dock Marsh/Future Mixed-Use

Dagenham Dock Industrial

ELT Phase 1

0m   500m   1000m

mmcité

mmcité is not only a supplier of high-quality street furniture; the company is also a partner to all those who want to create something special within public spaces. Mayors of cities of all sizes in the mountains or coastal areas, architect of small teams and large design institutions, construction companies of local or transnational importance – they address all of them with the aim of achieving the perfect project.

At the beginning, there is always a designer sketch, a mere intention. The strong team of experienced and educated professional create strong, high-quality products. Efficient functionality, careful processing and affordable cost are the main parameters that they monitor throughout the process. Modern design and distinctive expression represent a constant standard of mmcité.

They combine the very best materials, which are further continuously tested. They draw from two sources; on the one hand, from the years of their experience, on the other, from constant efforts to upgrade the materials. Only the best will pass through a sieve. Function, durability and of course price are what matter most. City public spaces are fascinating places where people meet history.

ID-Concepts

ID-Concepts is a young company in The Netherlands. Their products combine aesthetics, functionality, and innovation and thus offer the buyer the added value that is missing.

Quality, an emotion of their designs, sharpens the perception in the market and strengthens the identity. Together they will develop a product that reflects your corporate character and will delight the buyer in the long run.

Global, Arquitectura Paisagista, lda.

In 1997 Global, Arquitectura Paisagista is founded by João Gomes da Silva and Inês Norton, in Lisbon. Its subject is the investigation and development of concepts that arise from the ecological, social and

economic transformations in the Global Landscape. It is composed by a group of landscape architects and architects and regularly collaborates with other landscape architects, architects, engineers, designers and artists, constituting a strongly personalised and professionalised structure integrated in an interdisciplinary dynamic network. Global, Arquitectura Paisagista investigates, plans and develops projects within the scope of landscape architecture, including in its social subject the production of books and publications.

Slot

Slot is a an award-winning mid-sized architectural design firm based in Mexico City, working within the field of urbanism, architecture and interior design. Since its creation in 2008 by founding principals Juan Carlos Vidals and Moritz Melchert,

Slot has distinguished itself as a specialist in design and development of projects in all scales.

The architecture of Slot is characterised by contemporary design, which intends to establish an alternative way of producing architecture based on professionalism, analysis, planning, creativity, and the implementation of new technologies and multidisciplinary teamwork.

Slot is currently involved in a number of projects of different scales in Mexico, Europe and Asia. They include urban design proposals, single and collective dwelling projects, as well as public and cultural developments for governments and private investors. Just recently, Slot signed a contract of collaboration with the De Tao Masters Academy in Beijing, which includes teaching Master Studio Classes and further project collaborations.

Diana Cabeza

Diana Cabeza is an architect who studied at University of Belgrano, Buenos Aires, and graduated with honours from the Prilidiano Pueyrredon National School of Fine Arts.

Cabeza is a designer specialising in urban equipment, driven by re-thinking and designing urban elements and supports for community use in the public space. She is the principal designer and president of Estudio Cabeza Urban Elements for Public Spaces.

Her designs have been published in the following magazines: Domus, Abitare, The Plan, Nexus, Visions, Interior Design, Architectural Record, Summa + and Barzón as well as in books published abroad. Her designs have been licensed out to foreign companies in Europe and Latin America. Her urban pieces furnish public spaces in the cities of Buenos Aires, Córdoba, Puerto Madryn, Mendoza, Tokyo, Zurich, Paris, Washington DC, New York, Miami, Perth to name a few.

She obtained the Konex Platinum Award in industrial design and four Good Design Awards in 2012; the DARA Award in 2010; the first place in the National Competition of Urban Furniture for the city of Buenos Aires GCBA/SCA together with Wolfson/Heine in 2004; the ICFF Editors Award New York 2003 and the Konex Merit Award in 2002.

Ben Busche

Ben Busche was born in Munich, Germany in 1968 and studied Architecture and Urban Planning at the Technische Universität in Stuttgart, licencing in 1998. With the experience made at the Escuela Tecnica Superior de Arquitectura de Madrid, Spain in 1995 while realising a university interchange, he decided to move to Madrid after finishing his diploma, acquiring experience in several architecture offices collaborating in various projects of big scale, most of all hospitals. Simultanously, he carried on with his university career stuying for doctorate grade at the ETSAM. In 2004 he founded the office for architecture and design Brut Deluxe operating from two bases in Madrid and Munich. Brut Deluxe is focused on the investigation and creation of space and its atmospheric qualities. The projects oscillate between different scales of urban intervention: from ephemeral artistic installations to industrial design, construction design and urbanism.

Ignacio Ciocchini

Ignacio Ciocchini is an award-winning Industrial Designer specialised in street furniture products, urban design, and public space design. He is Vice President of Design at Bryant Park Corporation, 34th Street Partnership, and Chelsea Improvement Company, three leading Business Improvement Districts in New York City. As a design consultant, Ciocchini works with government agencies, transportation authorities, real estate developers, and architecture firms that have an interest in adding custom products to their projects. Ciocchini's designs have been included in exhibitions at the Copper Hewitt National Design Museum in New York City, The Autodesk Design Gallery in San Francisco, and The Guangzhou Design Week in China.

Minale Tattersfield

Minale Tattersfield Australia is a leading wayfinding design and branding

Minale Tattersfield
Sydney | London | Paris | Milan | Moscow | Brussels

agency. They are known for their ability to approach, analyse and understand the challenge first, and then allow the creative idea to evolve and develop. The work is strategically led, but creatively driven, aiming to energise and transform, to challenge convention, and inspire change.

They are part of the worldwide Minale Tattersfield Design Strategy Group, and have created designs that inspire and set benchmarks for their clients. They operate in an Australian context with the benefits of a global perspective.

Rojkind Arquitectos

© Rojkind Arquitectos

Rojkind Arquitectos is a Mexico City-based architectural firm with international practice focusing on tactical and experiential innovation. Founded in 2002 by Michel Rojkind, it is known for its innovative solutions in architecture and urban strategy. Gerardo Salinas joined the firm in 2010 as its first partner.

They use design thinking to cut across strategic fields looking to maximise project potential while maintaining attainability. A multinational team, Rojkind Arquitectos celebrates the collaborative nature of the architectural discipline and works side by side with experts in various fields who engage in the creative design process directed by an Adaptive Diagnostic Design or ADD system. ADD is a strategy, devised by the firm, to focus energy from research, local knowledge, multi ambit professionals and consultants, and other relevant tools in an effort to recognise, and work to achieve, every project's maximum potential. ADD is their way of thinking and their driver for everything from business strategy to design.

Rojkind Arquitectos has gained worldwide recognition for its award-winning projects including 2005 Architectural Record Top 10 Design Vanguard and 2011 Pro-México's business magazine "50 Names in the Global Creative Scene". Founding partner Michel Rojkind was named by Wallpaper* magazine in 2011 as one of the "150 Movers, Shakers and Makers that have Rocked the World in the Last 15 Years" and partner Gerardo Salinas as AIA's Denver Chapter "2008 Young Architect of the Year".

Thilo Frank

Thilo Frank (born 1978) lives and works in Berlin. His installations, sculptures and photo series lead the viewer into situations creating interactive physical dialogues. Physical phenomena and environments of our daily life are interpreted in a new context emphasising our perception of light, space and motion in a poetic and playful way. His works ask of a visitor to question one's relation to action in space and the consequences of one's action – an intimate experience. The viewer functions as a coproducer of the work, perceiving his or her relational aspect as a instrument of measurement. The cognitive aspect in its work is enacted by the viewer, who through his physical reaction to the work, reflects himself to it. Through his physical interaction the viewer completes the work.

Through the use of everyday technologies Thilo Frank explores the conventions of sight and movement and their visualisation. Examination produces a type of measuring standard through which even seemingly random events are viewed equally. Through his works, an attempt is made to illustrate the optical coefficient and with this to analyse the space.

MAIpublicspace

MAIpublicspace was set up by Miguel Costa as a research platform in order to develop new tools and interdisciplinary approaches from architecture, art, landscape architecture and other practices. These hybrid approaches and results seek to rethink the existing public spaces, creating more dynamic cities, increasing its urban quality and, consequently, of its inhabitant's lifestyles, reducing the asymmetries  between city centre and periphery, between places and non-places. The goal is to work between private, public and ephemeral projects that trigger situations, collective engagement, participation and new senses of appropriation and belonging.

rankinfraser landscape architecture

rankinfraser landscape architecture llp was formed in January 2008. The studio's approach is  guided by the following principles:

• The design emerges from an understanding and response to the unique context, possibilities and constraints of each site.
• The design emerges from an inclusive consultation approach involving the client, the design team, the end users and the public.
• The design emerges from the development of a strong spatial concept or set of governing principles.
• The quality of the design solution is paramount.
• The delivery of the design solution is vital and dependant on an understanding of detail design refinement and material possibilities,

rankinfraser makes landscape proposals which are innovative and practical, contextual and memorable, honest and durable, unobtrusive and unique, sustainable and original, value for money, pure and simple.

The Ground Studio

The Ground Studio is a Seoul-based architecture office in 2010. They want to create new thoughtful  relationship to the essence of various elements: funtion, architectural factors and the characteristics of a site.

The aim of their design is the simple integration of the function and aesthetic of material and spatial volume. All their practice have new architectural influence with the environment. This is the reason why The Ground Studio practice architectural studies.

Simone Bossi

Born in Italy in 1985, Simone Bossi studied Architecture at Politecnico of Milano and ETSA- Universidad de Sevilla. From 2009, he has been working in Italy, in The Netherlands and Switzerland for different offices. From 2011, he permanently works on his personal research and he investigates architecture through art, sculpture and photography.

Springtime

Springtime is a creative force that creates exciting, sustainable and paradigm-shifting products, brands  and experiences in the field of sustainable mobility, sports equipment, interior, public design, consumer electronics, juvenile products and brand development. They are based in Amsterdam and have their tentacles all over the globe. Springtime (1995) provides strategic consulting, design research, concept design and industrial design. They offer integrated design – a synergy of communication, graphic design, interface design, and product design. In order to do so, they work together closely with specialists in all fields. This way, they are able to provide their clients with exactly the innovation they need – from strategy to production. Brand, product and media are their ingredients for successful innovation.

Over the years, Springtime has been awarded with many design awards, both in The Netherlands and internationally. Springtime is member of the Dutch Designers Association (BNO).

Simon Watkinson

"Early in my development as an artist I realised a need to work outside of the gallery against the backdrop of the built environment, or rather entwined with it. Consequently my work evolves from a consideration of the physical and social characteristics of particular spaces or buildings. My approach could perhaps be described as the combination of interpretation and aspiration. The need to respond sensitively to the changing nature of the sites in which I work has also resulted in a wide range of approaches being undertaken. Recurring traits are beginning to emerge in my work, the use of lighting being one, but all are very carefully evolved from a direct engagement with the area." – Simon Watkinson

Keha3

Three men – Ville Jehe, Margus Triibmann and Tarmo Luisk – established OÜ Keha3 in 2009. Their idea was to create in Estonia the first design company managing entire production and sales chain. The areas of activity of Keha3 include design of own products; production and sales, design management and provision of design services. Currently the company employs six people.

The three men established Keha3 to make the world a better place. "We have ambitions, courage and experience to create big things: change your and our own fantasies, wishes and needs into tangible items." Many of their products are designed for outdoor use. If an item is capable to endure such conditions, it is capable to attend them also in much friendlier conditions. All their products are made in belief that they are the right things!

Javier Machimbarrena

"The objective of the creation of a product must be to permit the human being to satisfy a need and to arouse in him an emotion." Javier Machimbarrena (Bilbao, Spain, 1967), works as a product designer and industrial artist, creating and developing objects and spaces from the ones that are essentially form to the equipped with a high technological content.

He combines creative ability based on analysis with a conviction of productive feasibility, understanding Design in a global sense and conferring coherence to the object as a whole, to its cycle and to its surroundings.

In the last 15 years he has worked in projects involving the design of vehicles, furnishings, urban design, consumer products and exhibition spaces. He currently works in his own studio in Bilbao where he carries out projects for and in association with customers from various countries. Javier Machimbarrena also produces objects limited editions in his new experimental project called Industrial Art.

www.machimbarrenadesign.com

AHL Linghting Group

AHL Linghting Group specialises in producing LED pixel lighting. Servicing for engineering partners and lighting designers, AHL aims to become the best lighting manufacturer in global Media Façade industry. Quality-oriented, AHL always develops new technology, trying their best to become the world's first brand and the first choice manufacturer.

LYVR

LYVR is a design firm that is run by Lysbeth de Groot-de Vries (interior architect) and Johan de Groot (architect). Through the years they worked at several architectural firms and at the municipality of Groningen. Their designs range from fashion, furniture to public space and architecture.

Their mission is to create a more beautiful and uplifting surrounding for people to live in, also for people who cannot afford to pay for this. In that way they combine commercial and non-profit projects, at the moment located in The Netherlands, Eastern Europe, on the African continent and in India.

Atelier cité architecture

Atelier cité architecture is a workshop whose field of investigation crosses city territory and architecture. In very diverse professional practices (diagnosis, project teaching, research and publication), the different partners of cité architecture have always had the desire to develop global approaches and multidisciplinary territories and give great importance to the development of diagnostic to highlight the elements of identity of the place and build a contemporary project. Thinking about the long-term impacts of decisions on development, the environment and landscapes have always been the backdrop of all their work: the question of urban sprawl, the effects of "the all for car", the treatment of urban networks (travel, water, energy, infrastructure) and the identity of each place.

7N Architects

7N Architects is a design studio where architecture, urban design, masterplanning and landscape combine to bring a holistic approach to creating exceptional places.

Their team's strength lies in a comprehensive, ideas-led approach to unlocking the full potential of a project opportunity.

Their skills range from traditional architectural services to masterplanning, regeneration strategies and frameworks, design guidance, consultation, and public realm and civic design.

Their flexible, collaborative way of working allows them to blend their skills and adapt their approach to the clients' specific needs and visions for each project.

Designnobis

Designnobis is an award-winning design consultancy with a visionary approach on product design. Offering solutions to the world's contemporary problems by developing visionary and ecofriendly designs, it has been gathering attention globally since its foundation.

As a sustainable design centre, they introduce innovative solutions in product design, branding and space creation. They help their clients to differentiate in market by creating products and services with strategic value. Handling design services from research and concept development phase through prototyping and production, they aim to become a sustainable innovation centre.

Designnobis contribute to national and global industry by developing competitive products with high added value and they help to create sustainable development locally and globally. They provide R&D services for firms and add value to brand image in order to turn technology and inventions into products, services and good that are available to the consumer in the market. They offer total design services that are strategically and economically valuable while also differentiating brands.

Atelier 9.81

Founded in 2004, Atelier 9.81 is a town planner who works to develop and realise projects in the fields of architecture, town planning and stage design while adhering to an ethos of long-term development. Within this idea of long-term development, Atelier 9.81 is regularly called upon to participate in shows about accommodation and equipment throughout the whole of the French territory.

ATELIER 9.81
ARCHITECTURE URBANISME

Atelier 9.81 centres its attention on the conditions required for development of the quality of the answers. Methods, tools, procedures and systems. Not only does the studio focus on research, furniture, scenography, written production or explorations and links with disciplines covered by architecture and town planning, but also on mixed approaches that they use to the benefit of the architectural and town planning project.

SLA

SLA is an internationally renowned architecture firm that works with landscape, urban space and city planning. SLA creates modern, sustainable cities that

inspire community and diversity through innovative use of architecture, infrastructure, nature, design and technology.

BG STUDIO

BG STUDIO is a team of architects and designers based in Valencia. They like to think of it as a creative lab where they research the various aspects of design related to architecture and environment. Each project they take goes through an intense creative process where every

member of the team contributes a different point of view based on their different backgrounds.

RDG Dahlquist Art Studio

RDG Dahlquist Art Studio is a comprehensive design and fabrication studio within RDG Planning and Design, a nationally recognised multidisciplinary design firm with offices in Des Moines, Iowa; Omaha, Nebraska; Ames, Iowa; and Fort Myers, Florida. The work of David B. Dahlquist and RDG Dahlquist Art Studio is a thoughtful integration into public and private spaces, within building architecture and the landscape. The studio provides a unique process to the visual development of projects, combining insight with research and documentation. The studio produces individual artworks, as well as major site-specific installations, both structural and ornamental. Recognised for turning history and stories into meaningful, contemporary place-making installations, many different elements are "orchestrated" in a way in which the public is welcomed into and moved by the experience. Paying close attention to the quality of how things are made, the studio works collaboratively with architects and engineers to make sure that structurally what is proposed will function as intended and will last. In the process, RDG Dahlquist Art Studio has created memorable destinations and exciting new places.

Pensa

Pensa is a strategic design consultancy with a track record of developing successful products. Pensa's

product designs connect with consumers because they are rooted in an understanding of people, the products they use, and the contexts in which they use them. Pensa's team of dedicated, talented individuals each has expertise in multiple fields and come from a variety of backgrounds. Their experience spans a broad range of industries including housewares, home healthcare products, medical devices, hardware tools, childcare products, office furniture and consumer package goods.

Alexander Lotersztain

Alexander Lotersztain was born in Buenos Aires, Argentina in 1977; he graduated from Design at Griffith University QCA in 2000. He is director of Derlot Pty. Ltd., a multidisciplinary studio focusing on projects including product, furniture, branding, hotel design, interior design and art direction with clients both nationally and internationally.

Derlot Editions is co-brand of derlot and produces a range of Australian made furniture and lighting for the contract and domestic markets and distributed worldwide.

Mr. Lotersztain won the Inaugural Queensland Premier's Smart State Designer of the Year Fellowship Award 2010, and recently returned to judge the award for 2011. He was named one of 100 most influential top designer worldwide in &fork by Phaidon, top 10 most influential faces in Design by Scene Design Quarterly 2007 and top 10 of 100 Young Brightest Australian Achievers Bayer/Bulletin Award. He has won many awards in both product and interior design and his work has appeared in design journals around the world. Alexander is also part of the "Smart State Design Council" for the Queensland Government in Australia, drafting the Smart State Design Strategy for 2020.

www.derloteditions.com

UNO+UNA

Plácido Piña and Rafael Calventi, architects with a strong and successful career, led the team with the collaboration of Architect Esteban González and Designer Noemí Zaro (UNO+UNA) in the urban design and landscaping and the Architect Gina Calventi as lighting architect advisor.

arquitectura
interiorismo
UNO
+UNA
hola@unomasuna.com
www.unomasunablog.com

ANNABAU Architecture and Landscape

ANNABAU
architektur und landschaft

choriner str. 55
10435 berlin
mail@annabau.com
www.annabau.com

ANNABAU is a young interdisciplinary office for architecture and landscape architecture based in Berlin. Office partners are landscape architect Sofia Petersson and architect Moritz Schloten.

ANNABAU is an office planning projects with a high demand on design and on spatial solution issues. High-quality execution, compliance with budget and schedule as well as flexibility are important to the company. In addition to architecture projects in the private and public sectors ANNABAU designs gardens, parks, public squares and playgrounds. Extensive experience in energy planning and energy efficiency makes sustainability an integral part of the projects.

Tonkin Liu

Tonkin Liu is an award-winning architectural practice, whose work encompasses architecture, art and landscape. They offer forward-thinking clients a design that is finely tuned to the place it is sited, the people who will occupy it, and the culture that surrounds it at the time.

This emphatic search for new beginnings is set out in their book "Asking, Looking, Playing, Making", published in 1999. The unique storytelling methodology searches for archetypes that will inform the process of design from inception to completion, giving the project a lasting resonance.

Each project embodies the relationship to nature. Some projects celebrate changing weather and seasons, some evoke the power of nature as symbols, whilst others emulate form and performance, using lessons in nature to inspire pioneering construction techniques. Their preoccupation with nature informs the design process, whether through biomimicry or by using the elements nature generously gives human beings for free.

They are interested in doing what they have not done before and their aim is always the same, to satisfy the mind and touch the heart.

B|D landscape architects

B|D landscape architects have outstanding expertise in the field of public realm design and a growing reputation for contemporary landscape architecture, urban design and space-making.

B|D landscape architects is a dynamic, energetic and dedicated firm committed to the integration of sustainable design to deliver amazing places. Their approach combines contemporary sustainable design with creative ecology looking to glean the unique genius loci from each place.

They believe in creating extraordinary effects with ordinary things to deliver meaningful and much loved public spaces. They relish collaborating in multidisciplinary teams pushing the boundaries of contemporary design using technological innovation and sustainable materials.

Ian McChesney

Ian McChesney works as an independent architect, designer and sculptor. Commissions include "Out of the Strong Came Forth Sweetness" in London's Angel Building, rotating wind shelters for Blackpool's promenade, an award-winning park pavilion in Preston and a range of small batch-produced lamps. He has recently completed Arrival and Departure for the grounds of Plymouth University in Devon.

Microarquitectura

microarquitectura

Microarquitectura was founded in 1995, addressing its activity to design, construction and commercialisation of urban furniture, especially kiosks and pergolas, always focusing on projects development. They make normalised catalog proposals or special projects and customisations. They normally carry out singular projects of signing or sculpting to create urban furniture always as a project. They pursue quality and design and this is proved in their every action, as well as in their active participation in prestigious associations like ADI FAD or the European project "Design for all", where they are part as promoter of the Foundation. They usually work with and for the Administration, with a strong concern for sustainability, design and quality. Their designs are largely and efficaciously proved. Nowadays, Microarquitectura's products catalog includes three divisions: urban furniture, shelters and recreational areas, covering a large market segment that enables to offer the Administration and private clients customised solutions for each case.

John Potter

John has 22 years' professional experience as a Landscape Architect, and is a Principal in the Auckland Office. John joined Boffa Miskell Auckland in 2001 after moving from the UK, where he had gained 12 years' experience as a Landscape Architect working in both public and private sectors. Prior to leaving the UK John was an Associate of a medium sized design-led Landscape Consultancy focusing on landscape design with supporting Planning and Ecology teams.

John has extensive experience of design, project management and implementation of a wide range of projects operating in a multidisciplinary environment, with particular emphasis on large-scale commercial, retail, healthcare, urban improvement and residential schemes. In addition, he has expertise in a variety of project procurement processes including traditional contract, design and build and partnership frameworks. He is a current member of the Auckland Council Urban Design Panel.

Burns + Nice

Burns + Nice is a specialist urban design, landscape architecture, environmental and transport planning consultancy with extensive experience of a wide range of urban planning and design projects undertaken in the UK, Europe, the Middle and Far East.

They deliver solutions that provide social, cultural and economic gain to local communities and businesses based on consultation, focused strategies, innovative designs and achievable programmes for implementation. Burns + Nice recognise the need to protect the environment and to use resources in a way that provides for future generations by promoting sustainable design and procurement in both the consultancy services that they provide and the day-to-day management of the company.

Barbara Grygutis

Barbara Grygutis is an award-winning, U.S.-based sculptor and public artist whose work is in permanent public art collections throughout the United States and beyond, including Miami, Florida; Philadelphia, Pennsylvania; New York City; Washington DC; Denver, Colorado; Seattle, Washington; Columbus, Ohio; St. Paul, Minnesota; Calgary, Alberta, Canada; Phoenix, Arizona, and in Tucson, Arizona, where she lives and works. Light, natural daylight and artificial light, is integral to much of Barbara Grygutis' work. With her formal training in architecture and fine art and an on-going love of historic preservation, public art combines her passions. The artist's interest in nature is evident, and though the natural world provides the genesis of her working vocabulary, such forms are significantly altered through concept development, scale, and materials. Her works of art create a place of reflection, where the beauty of the natural world can be seen in the built environment.

UPI-2M

UPI2M
architecture | structure | design | consulting

UPI-2M, design company from Zagreb, Croatia is led by four partners. The main company's business operations include: architectural design, structural design, consulting services and graphic design. The company also practices urban planning, engineering services and supervision.

UPI-2M is dedicated to provide clients with the core of a full design package – architecture & structure, starting from the very beginning of each project. By integrating those two elements as a whole, along with other disciplines, it explores and conceives unique and integral project solutions. Thanks to such holistic approach, the company has successfully met quality expectations to numerous project assignments and conceived many structural and architectural solutions, always exploring and implementing contemporary methods and contemporary approaches.

scape landscape architects

The office scape landscape architects was founded in 2001 by Matthias Funk, Hiltrud M. Lintel and Rainer Sachse in Duesseldorf. For their clients they are working with a young and motivated team, in cooperation with city planners, architects, ecologists and communication designers mainly on object-planning designs of urban open spaces. The current projects range from master development plans for entire districts over conceptual designs for parks, pedestrian streets and squares, through to detailed planning of their own street furniture systems.

Omos Ltd.

omos

Leading the way in street furniture design and innovation, Omos manufactures a comprehensive range of contemporary products including litterbins, recycling bins, benches, seats, tree planters and grilles, bollards, ashtrays, picnic sets, cycle stands and shelters.

Omos has been designing and manufacturing street furniture in Ireland since 1996. Their products have received numerous awards including The Glen Dimplex Grand Prix, The ICAD Silver Bell, The Louth County Council and Carlow County Council Street Furniture Awards.

Design and engineering go hand in hand in every Omos product. Their clean uncluttered aesthetic demands that the product's workings and structure be part of this aesthetic. Omos achieves this through a parallel approach to design; from conception, evaluating design and engineering against the functional requirements of the product. In addition to their contemporary product range, Omos offers a bespoke design and manufacture service. With many years' experience in bringing ideas from conception to production, they offer an effective means of producing ideas.

matali crasset

Matali Crasset is by training an industrial designer, a graduate of the Ateliers – E.N.S.C.I. (Workshops – National Higher School of Industrial Design). At the beginning of 2000, after her initial experience with Denis Santachiara Italy and with Philippe Starck in France, she set up her own studio in Paris called matali crasset productions in a renovated former printing firm in the heart of Belleville. It is there, with the coming and going of children and neighbours that she dreams up her projects.

She considers design to be research, working from an off-centred position allowing to both serve daily routines and trace future scenarios. With both a knowledgeable and naive view of the world, she questions the obviousness of codes so as to facilitate her breaking these bonds. Like her symbolic work, focused on hospitality, "Quand Jim monte à Paris" (When Jim goes up to Paris), is based on a mere visual and conscious perception which she invents another relation to the everyday space and objects. Her proposals are never towards a simple improvement of what already exists but, without rushing things, to develop typologies structured around principles such as modularity, the principle of an interlacing network, etc. Her work revolves around searching for new coordination processes and formulating new logics in life. She defines this search as an accompaniment towards the contemporary.

Ruedi Baur

Designer, born in 1956, French and Swiss nationalities, graduated in graphic design from the Zürich School of Applied Arts. After having created BBV (Lyon - Milan- Zürich) in 1983, he set up in 1989 Integral concept, presently constituted of five independant partners studios being able to intervene jointly on any crossdisciplinary project. Since 1989 in Paris, 2002 in Zurich and 2007 in Berlin, Intégral Ruedi Baur has been working on two- and three-dimensional projects within the different fields of visual communication: identity, orientation and information programmes, exhibition design, and urban design. Between 1989 and 1994, he coordinated the department of design "information space" of the Beaux-Arts school of Lyon where he organises between 1994 and 1996 a third cycle based on "civic and design spaces" theme. In 1995, he became professor at the Hochschule für Grafik und Buchkunst of Leipzig, at which he was nominated rector from 1997 until 2000. He created there in 1999 the Interdisciplinary Design Institute (2id). Between 2004 and 2012, he was in charge of the Design Institue of the Hochschule für Gestaltung und Kunst der Stadt Zürich (HGKZ). Since 2007, he teaches at the "Ecole Nationale Supérieure des Arts Décoratifs de Paris" (ENSAD). In 2011, he created two research institutes: Design4context and Civic City. He started to manage the "Certificate of Advanced Studies – CAS Civic Design" at the Head in Geneva.

Rehwaldt Landschaftsarchitekten

In 1993 the office was founded by Dresden landscape architect Till Rehwaldt.

Main sectors of their scope of work are public open space planning, recreation and leisure facilities and urban planning.
They are working on miscellaneous thematically and regionally diverse projects.

For them, taking part in design competitions is an ambitious recurring challenge. Most of their projects have been generated through this kind of planning process.

Working fields:
Open space planning, object planning, conceptual open space planning, urban planning, revitalisation of housing areas, remodelling of mining landscapes.

后 记

　　本书的编写离不开各位设计师和摄影师的帮助，正是有了他们专业而负责的工作态度，才有了本书的顺利出版。参与本书的编写人员有：

matali crasset, matali crasset productions, mmcité, Minale Tattersfield, Hans Gerber, Global, Arquitectura Paisagista lda., ID-Concepts, Ruedi Baur, Diana Cabeza, Martín Wolfson, Leandro Heine, Ben Busche, Miguel de Guzman, Brut Deluxe, Ignacio Ciocchini, Marco Castro, rojkind arquitectos, Jaime Navarro, SLOT, Thilo Frank, ANNABAU Architecture and Landscape, Miguel Costa, 7N Architects, rankinfraser landscape architecture, Dave Morris, Ian McChesney, Simon Watkinson, Sean Conboy, simone bossi, Springtime, b:d landscape architects ltd, Pensa, Keha3, Simon Watkinson, Javier Machimbarrena, AHL Linghting Group, LYVR, Lysbeth de Groot-de Vries, Johan de Groot, Atelier Cité Architecture, The ground Studio, Designnobis, Atelier 9.81, Julien Lanoo, Barbara Grygutis, Margaret Kirkpatrick, BG STUDIO, Javier Guijarro Tortosa, David B. Dahlquist, Don Scandrett, UPI-2M, Vanja Solin, Davor Konjikusic, Studio Blagec, Studio Derlot, Tonkin Liu, Florian Groehn, Rehwaldt Landschaftsarchitekten, Rehwaldt LA, Uno+ Una. Arquitectura e Interiorismo Architecture, Omos, Microarquitectura, Marc Guillen, Boffa Miskell, scape landscape architects, SLA, BURNS +NICE, Transport for London, Tim Crocker.

ACKNOWLEDGEMENTS

We would like to thank everyone involved in the production of this book, especially all the artists, designers, architects and photographers for their kind permission to publish their works. We are also very grateful to many other people whose names do not appear on the credits but who provided assistance and support. We highly appreciate the contribution of images, ideas, and concepts and thank them for allowing their creativity to be shared with readers around the world.